Constructing Scientific Psychology
Karl Lashley's Mind-Brain Debates

Constructing Scientific Pschology is the first full-scale interpretation of the life and work of the major American neuropsychologist Karl Lashley that sets Lashley's creation of a laboratory-centered, decisively materialistic science of brain and behavior in its scientific and social contexts. The book sets Lashley's neuropsychology at the heart of two controversies that polarized the sciences of mind and brain in the U.S. in the first half of the twentieth century. The first concerned the place of "consciousness" and "free will" in the nervous system; the second concerned the relative roles of "nature" and "nurture" in shaping behavior and intelligence. Drawing on Lashley's extensive unpublished correspondence as well as his published scientific papers, the book argues that his neuropsychology and his experimental practice were both scientific and political tools. Despite his attempt to create a "pure," fact-driven science, free of social applications and theoretical presuppositions, Lashley's work was at once a scientific answer to the problem of brain and behavior, a hereditarian answer to the problem of intelligence, and a nativist and deeply conservative answer to the problem of racial integration in American society.

Nadine M. Weidman is a postdoctoral research scholar and tutorial assistant in the Department of the History of Science at Harvard University, and an instructor in the Harvard Extension School. She has previously taught history of science at Cornell, Harvard, and Worcester Polytechnic Institute. Dr. Weidman is editor of the *Cheiron Newsletter* and has contributed articles and book reviews to the *Journal of the History of the Behavioral Sciences, Journal of the History of Biology, History of Psychology, American National Biography,* and *Isis.*

T0275704

Cambridge Studies in the History of Psychology

MITCHELL G. ASH AND WILLIAM R. WOODWARD

This series provides a publishing forum for outstanding scholarly work in the history of psychology. The creation of the series reflects a growing concentration in this area by historians and philosophers of science, intellectual and cultural historians, and psychologists interested in historical and theoretical issues.

The series is open both to manuscripts dealing with the history of psychological theory and research and to work focusing on the varied social, cultural, and institutional contexts and impacts of psychology. Writing about psychological thinking and research of any period will be considered. In addition to innovative treatments of traditional topics in the field, the editors particularly welcome work that breaks new ground by offering historical considerations of issues such as the linkages of academic and applied psychology with other fields, for example, psychiatry, anthropology, sociology, and psychoanalysis; international, intercultural, or gender-specific differences in psychological theory and research; or the history of psychological research practices. The series will include both single-authored monographs and occasional coherently defined, rigorously edited essay collections.

Constructing Scientific Psychology

Karl Lashley's Mind-Brain Debates

Nadine M. Weidman

CAMBRIDGE
UNIVERSITY PRESS

CAMBRIDGE UNIVERSITY PRESS
Cambridge, New York, Melbourne, Madrid, Cape Town, Singapore, São Paulo

Cambridge University Press
The Edinburgh Building, Cambridge CB2 2RU, UK

Published in the United States of America by Cambridge University Press, New York

www.cambridge.org
Information on this title: www.cambridge.org/9780521621625

First published 1999
This digitally printed first paperback version 2006

A catalogue record for this publication is available from the British Library

Library of Congress Cataloguing in Publication data
Weidman, Nadine M., 1966–
Constructing scientific psychology : Karl Lashley's mind-brain
debates / Nadine M. Weidman.
p. cm. – (Cambridge studies in the history of psychology)
Originally presented as the author's thesis (doctoral) – Cornell
University, early 1990s.
Includes bibliographical references.
ISBN 0-521-62162-3 (hardcover)
1. Lashley, Karl S. (Karl Spencer), 1890–1958. 2. Psychologists –
United States – Biography. 3. Neuropsychology – History. I. Title.
II. Series.
BF109.L37W45   1999
150'.92 – dc 21   98-26461

ISBN-13 978-0-521-62162-5 hardback
ISBN-10 0-521-62162-3 hardback

ISBN-13 978-0-521-02777-9 paperback
ISBN-10 0-521-02777-2 paperback

For my family
with love and gratitude

But who shall parcel out
His intellect by geometric rules,
Split like a province into round and square?
Who knows the individual hour in which
His habits were first sown, even as a seed?
Who that shall point as with a wand and say
"This portion of the river of my mind
Came from yon fountain?" Thou, my friend, art one
More deeply read in thy own thoughts; to thee
Science appears but what in truth she is,
Not as our glory and our absolute boast,
But as a succedaneum, and a prop
To our infirmity. No officious slave
Art thou of that false secondary power
By which we multiply distinctions, then
Deem that our puny boundaries are things
That we perceive, and not that we have made.
To thee, unblinded by these formal arts,
The unity of all hath been revealed,
And thou will doubt with me, less aptly skilled
Than many are to range the faculties
In scale and order, class the cabinet
Of their sensations, and in voluble phrase
Run through the history and birth of each
As of a single independent thing.
Hard task, vain hope to analyze the mind,
If each most obvious and particular thought,
Not in a mystical and idle sense,
But in the words of Reason deeply weighed,
Hath no beginning.

William Wordsworth
The Prelude, Book II,
lines 203–232

Contents

Preface

This book began as my dissertation in the Department of Science and Technology Studies at Cornell University in the early 1990s. It was an exciting context in which to be studying the history of science: a radical skepticism toward scientific authority and scientific truth had begun to be taken as the sine qua non of serious historical scholarship. Science, my colleagues and I were taught, was thoroughly informed by society; scientific theories and practices were products of culture, not nature; laboratory experimentation was an elaborate ritual ripe for anthropological analysis. In our seminars and discussions, there was the pervasive sense that we were breaking with tradition, riding the wave of a revolutionary new approach to the field.

In part what made these new ideas so exciting was the debate that swirled around them. Not everyone at Cornell was a "social constructivist," and the controversy about the relationship between science and culture was heated and ongoing. As a graduate student, I found it impossible not to define my work somehow in relation to the arguments I observed and participated in.

While I became persuaded of the usefulness of a social constructivist perspective in doing history of science, I was also acutely aware of the criticisms brought against it. I began to think that the most important contribution I could make to the debates would be to demonstrate the range and power of social constructivist ideas: to show how social constructivism could account for the hardest cases in the history of science. I did not intend to treat social constructivism as a scientific hypothesis, to be tried and tested on historical subjects and discarded if found lacking. Rather, I believed it was a way of looking at the world, a way of making sense of science, its history, and its relationship to society. I was not looking to test the "truthfulness" of social constructivism; rather, I planned to show how useful a tool it could be for the historian. It was with this deliberate purpose in mind that I chose the subject of this book.

One day, while I was casting about for a dissertation topic in the history of the brain and behavioral sciences, I happened, quite by accident, upon a description of the comparative psychologist Karl Lashley. A pure scientist, he was called; politically and socially neutral; dedicated to the facts that emerged from his painstaking laboratory work. A cursory review of his scientific papers revealed the accuracy of this description; and the more I became acquainted with his writings, the more amply it was confirmed. If anything was a reflection of nature, not a construction of culture, it was Lashley's science: surely here was one of those hard cases I was looking for. I therefore set myself a very conscious challenge: to show that the social constructivist methods and approaches I had adopted could make sense of Lashley's life and work.

I wrote this book with another purpose in mind as well, though this one emerged only gradually during the course of my research. I was a newcomer to the history of the mind, brain, and behavioral sciences, and I was always looking for books that would put it all together for me, that would draw the big picture. I was always looking for books that would discuss the relationships *between* scientific fields and specialties, so that I could understand Lashley's position with regard to the many disciplines on which his work drew. When I read something about the rise of behaviorism, for example, I wanted to know how it was situated not only with regard to the history of comparative psychology, but how it fit more broadly within psychology as a whole, including social psychology and psychoanalysis, as well as within the other social sciences, and how it was related to the history of the biological fields, physiology, neurology, evolution, genetics. I wanted books that would draw a cognitive map of the life and human sciences in the twentieth century. I did find some books and articles that did this, that did not remain firmly rooted in the rise of one discipline or professional specialty. My debts to their authors are recorded in the pages that follow. As my project developed, I wrote very much with the hope that my book too would be one of those that attempted to draw the big picture.

The structure of my argument should perhaps be made plain at the outset. This book constructs an image of Lashley and then, at the end, deconstructs it. It first builds up a case for the "neutrality" of Lashley's science, showing the resources and strategies that Lashley used to project an image of pure science: an opposition to theory as well as applications; a sole reliance on the "facts" of brain function; and a firm commitment to the hereditary determination of behavior. I will argue that Lashley used his hereditarianism to bolster his claim to purity and neutrality. At the end of the book, I will show that Lashley's very disinterestedness was itself a political standpoint, and, by relying on his private correspondence, reveal the specific political beliefs and social ideals bound up with his scientific work.

Cambridge, Massachusetts
November 1997

Acknowledgments

Several institutions and many individuals made this work both possible and enjoyable. I am grateful to the Department of Science and Technology Studies and its predecessor, the Program in History and Philosophy of Science, at Cornell University, for their support of my dissertation research, out of which this book developed. From the beginning, my graduate committee provided invaluable constructive criticism and encouragement. Will Provine generously shared with me his immense knowledge of the history of biology and the riches of his personal library; he provided me with a model of historical scholarship while at the same time always giving me the freedom to express my own ideas. Peter Dear taught me how to be a historian; his incisive comments on everything I wrote – particularly his question, "But what was at stake?" – helped me immeasurably to clarify my thinking. Larry Moore's insights into the connections between American science and American culture were crucial in helping me to formulate my argument. I thank the members of the department for creating an exciting intellectual community, especially Sheila Jasanoff, Margaret Rossiter, Trevor Pinch, and L. Pearce Williams.

A year in MIT's Program in Science, Technology and Society as a Mellon Postdoctoral Fellow gave me the opportunity for rethinking and reimagining necessary for transforming a dissertation into a book. I owe thanks to all the members of the program, especially Lily Kay for invaluable constructive criticism, ideas, and suggestions, and Evelyn Fox Keller for much stimulating conversation. I have found an intellectual home in the Department of the History of Science at Harvard University, and I am grateful for the support and interest of its faculty, graduate students, and postdoctoral fellows. I owe special thanks to Everett Mendelsohn, for insightful suggestions and for intellectual and editorial support; and to Anne Harrington, whose pioneering work in the history of the mind, brain, and behavioral sciences has been a model for my own.

I am grateful to the organizers of and participants in five summer courses in history of biology at the Marine Biological Laboratory at Woods Hole (1987, 1989, 1991, 1992, and 1993), whose conversation has shaped my understanding of the field. I have also benefitted enormously from audience response to my talks on Lashley, especially at Cheiron (June 1994 and 1995), at MIT (March 1995), at the University of New Hampshire (February 1996), and at The Johns Hopkins University (October 1997). Harvard's Mind, Brain and Behavior History Group and Life Sciences Research Group have provided wonderfully stimulating forums for the discussion of ideas.

The staff of the Department of the History of Science at Harvard has been thoroughly kind, helpful, and supportive. I owe thanks especially to Meg Alexander, Billie Jo Joy, Betsy Smith, and Jean Titilah.

I thank the following archivists for their helpful assistance and for permission to quote from their collections: at the Archives of the History of American Psychology, Sharon Ochsenhirt, Marion White McPherson, John A. Popplestone, John Miller, and Sue Mann; at the Alan Mason Chesney Medical Archives at the Johns Hopkins Medical Institutions, Arian Ravanbakhsh and William R. Day Jr.; at the Ferdinand Hamburger Jr. Archives at The Johns Hopkins University, James Stimpert; at the Yerkes Regional Primate Research Center, Nellie Johns and Frederick King; at the American Philosophical Society Library, Scott DeHaven; at the University of Florida at Gainesville, Carl Van Ness; and the staff of the libraries at Cornell and Harvard. I am grateful to Christina Schiller Schlusemeyer for permission to quote from the papers of Karl Lashley at the University of Florida at Gainesville. I also thank the Clinical Psychology Publishing Company for permission to reprint parts of my article "Mental Testing and Machine Intelligence" (*Journal of the History of the Behavioral Sciences* 30, April 1994, pp. 162–180); Kluwer Academic Publishers, Inc., for permission to reprint parts of "Psychobiology, Progressivism and the Anti-Progressive Tradition" (*Journal of the History of Biology* 29, 1996, pp. 267–308); and John Wiley and Sons, Inc., to reprint parts of "Heredity, Intelligence and Neuropsychology; Or, Why *The Bell Curve* is Good Science" (*Journal of the History of the Behavioral Sciences* 33, Spring 1997, pp. 141–144). I am grateful for the editorial support I received from these journals. I am also grateful to the McGraw-Hill Book Company and the University of Chicago Press for permission to use illustrations from Lashley's work.

Many people showed me that scholarship need not be the lonely uphill battle of atomized individuals, but a truly collective enterprise. I am indebted to Leslie Barber, Bonnie Blustein, Roberta Brawer, Andrea Burrows, John Carson, Deb Coon, Michael Aaron Dennis, Anne Fausto-Sterling, J. Nadine Gelberg, Richard Held, Lillian Isacks, Dvora Kamrat-Lang, Stephanie Kenen, Petra Lucht, Jane Maienschein, Tom McGowan, Richard Noll, Tamara Ohr-Campbell, Robert Olby, Mary Parlee, Miranda Paton, Hans Pols, Karen Rader, Robert J. Richards, Rachael Rosner, Franz Samelson, Judy Schloegel, Irina Sirotkina, Betty Smocovitis, Mike Sokal, Susan Spath, Kamal and S. N. Sridhar, Frank Sulloway, and Marga Vicedo. Allison Hart Lengyel's perspective on history and culture is a continuing source of illumination. Sara Tjossem's incisive commentary, wit, and wisdom have benefitted me beyond measure. Darryl Bruce generously shared his work with me and offered constructive criticism. Donald Dewsbury gave me the benefit of his commentary and his far-reaching knowledge of history and of psychology, generously shared with me his ideas about Lashley, and patiently listened to my interpretations, even when they disagreed with his own. Sharon Kingsland encouraged my development as a scholar at every step and helped me to understand the history of the life sciences in new and productive ways.

I was fortunate to have the perceptive and detailed commentary of two very wise readers for Cambridge University Press: John Burnham and A. Edward Manier. Mitchell Ash and Roger Smith also provided invaluable suggestions on the whole manuscript. I am grateful for editorial support from Mitchell Ash, William Woodward, and Alex Holzman.

I thank my teachers at Bryn Mawr College, especially Mabel Lang, Paul Grobstein, Michael Krausz, and the late Jane Oppenheimer, for not only being patient with but also encouraging my diverse interests, and for their inspiring teaching.

Words are inadequate to express my deep gratitude for the love, encouragement, and interest that are shown me every day by the members of my family, and that sustain my engagement with my work. It is to them that I dedicate this book. I thank my mother, Bette Weidman, for talking over ideas with me and braving the wilds of Akron and Atlanta; my sister, Amanda Weidman, for sharing her knowledge of critical theory and anthropology; my father, Burton Weidman, for discussions of science and medicine; my grandmother, Ethel Statsky, for critical and careful reading; my parents-in-law, Martha and Joe Ferrari, Sr., and Mary and Carl Blais, for support, interest, and understanding; and my husband, Joe Ferrari, Jr., for listening to me thoughtfully, asking me hard questions, and for his humor and companionship.

Many people helped me to shape my ideas about Lashley. But the responsibility for the story told in these pages is mine alone.

Abbreviations

AHAP   Archives of the History of American Psychology
APS   American Philosophical Society Library
JCN   *Journal of Comparative Neurology*
JHBS   *Journal of the History of the Behavioral Sciences*
JHU   Johns Hopkins Medical Institutions, Alan Mason Chesney Medical
       Archives
NRC   National Research Council
UFG   University of Florida at Gainesville
Yerkes Yerkes Regional Primate Research Center

Introduction

The question at the heart of this study has a history of more than two thousand years; while it has invited solution after solution, it never seems to get solved. How is "the marvellous phenomenon of the mind" produced from "the enigmatic three-pound mass of tissue known as the brain"?[1] How can chemicals, cells, electrical signals – in short, matter – give rise to our consciousness, our thoughts, dreams, hopes and fears? How can two such different categories of existence bear any relationship to each other, much less be born, live, and die together?

Why the mind-body problem has remained so peculiarly intransigent despite repeated attacks is in itself a question worthy of consideration; that it continues to invite attacks cannot be disputed. In his recent book *Consciousness Explained,* Daniel Dennett asked, "how could the brain be the seat of consciousness?" and then proceeded to give the following answer:

> It turns out that the way to imagine this is to think of the brain as a computer of sorts. The concepts of computer science provide the crutches of imagination we need if we are to stumble across the *terra incognita* between our phenomenology as we know it by "introspection" and our brains as science reveals them to us.[2]

Meanwhile, John Searle has announced his own "simple" solution to the mind-body problem, one he hopes will "put the final nail in the coffin of the theory that the mind is a computer program."[3] Lest we conclude that the problem is today mainly for philosophers, the biologist Gerald Edelman has written a trilogy describing what he calls the only "biological" solution to the mind-body problem.[4] And the neurologist Oliver Sacks has called for a theory of brain and mind that does justice to the dignity of the soul.[5]

In this study, I have chosen to examine a brief segment of this long history in which several different solutions to the mind-body problem were proposed by

---

[1] Steven Levy, "Dr. Edelman's Brain," *The New Yorker,* May 2, 1994: 62.
[2] Daniel Dennett, *Consciousness Explained* (Boston: Little, Brown, 1991), 433.
[3] John Searle, *The Rediscovery of the Mind* (Cambridge: MIT Press, 1992), 1, xi.
[4] Levy, 62.
[5] Oliver Sacks, "Neurology and the Soul," *The New York Review of Books,* November 22, 1990: 50.

psychologists and biologists. The controversies that erupted among them took place between 1910 and 1955 in various American universities, and have one individual at their focal point: the American comparative psychologist Karl Spencer Lashley.[6]

Lashley's solution to the mind-body problem is one that many people today would probably find congenial: the mind is nothing but the brain; there is no immaterial soul; thoughts and feelings can be explained entirely in terms of neurons and chemicals. But I have not chosen to concentrate on Lashley simply because of the similarity of his ideas to our own. Rather, his story shows us that much more is at stake in the solution to this persistent problem than a straightforward desire to figure out the relationship between mind and brain. The solution Lashley proposed was based on hard experimental evidence: his theories rested on what he believed were the facts. But his positions in the different controversies, his ideas about how the brain worked, were not solely the products of his laboratory. They were also the result of his position within his relatively young profession – psychology – and of his political and social ideals. The story of Lashley's life and work offers a casestudy in the continuous interconnections between scientific inquiry, political ideology, and social context.

To demonstrate the relationship between Lashley's context – his beliefs about the world, his hopes for society and for his profession – and his position in these debates is, then, one of the main tasks I have set myself. Lashley made a point throughout his career of constructing a radical separation between his science and his politics. For him, the problem of mind – indeed, all of science – was independent of culture, of social interests, of moral implications. One of my purposes is to show that the science and its context were inseparable. In doing so, I argue that we must begin to broaden the focus of our interest in the mind-body problem: not only to look for an answer, but also to ask why so many people for such a long time have been interested in an answer, and to ask why their answers look the way they do. We must not simply accept one solution or another, but ask what historical conditions made such solutions possible. We should try to understand not only the facts, but also how the facts get constructed.

Though his name is little known today, especially among nonscientists, Lashley was the premier American brain researcher of the first half of the twentieth century. In his lifetime, he was celebrated as the author of an innovative and influential theory of brain function: the idea that the brain functions as a whole, not as a congeries of discrete capacities. This conception rested on a series of

[6] The discipline of psychology before the Second World War was focused on questions relating to the mind, its functions and capacities and its relation to the body, rather then on therapeutics as it is today. See Michael M. Sokal, "The Gestalt Psychologists in Behaviorist America," *American Historical Review* 89 (1984): 1241.

Karl S. Lashley (photograph probably taken around 1950). (Reprinted from *The Neuropsy-chology of Lashley*. Used with the permission of the publisher.)

experiments that Lashley conducted on the brains of rats, experiments so meticu-lous that they served as a model of objectivity for a generation of investigators.

Revered as he was in life, since his death in 1958 Lashley's stature has only increased, especially among the scientists who followed his lead. Although cer-tain of his specific findings have been disproved by later research, his general approach to the problems of brain, mind and behavior has been hailed as pioneer-ing and revolutionary, a definitive break from the entrenched and obscurantist traditions in early twentieth century American psychology, and a harbinger of post-1950s developments in the sciences of brain and mind.

In his history of the cognitive revolution, for example, the psychologist/ historian Howard Gardner names Lashley as the iconoclast and visionary who helped to make possible the interdisciplinary project of cognitive science, a sci-ence that combined approaches from psychology, linguistics, philosophy, neurol-ogy, and artificial intelligence in order to solve the mind-body problem. By jettisoning the old-fashioned models of neural function and turning toward a scientific explanation of higher mental processes, "Lashley helped set the stage for a cognitive science approach to behavior and thought."[7]

Likewise, in a volume celebrating the inception of neuroscience, an inter-disciplinary conglomeration of neurophysiology, neuroanatomy, and neuro-chemistry, Lashley's name is repeatedly invoked by such luminaries in the field as Roger Sperry, Alexander Luria, Wilder Penfield, and Ralph Gerard. Lashley, they agreed, pioneered the idea that the highest and most complex mental processes could be addressed by the tools of biological science. Lashley made the "neural correlates of conscious experience" a worthy and reputable problem for science; he showed his followers that the strengths of physiological psychology could be directed toward the mysteries of thought, feeling, and memory. The neuroscientist Eliot Stellar put it most vividly when he said that Lashley had stood at the crossroads of psychology and neurology "just as the traffic was beginning to roar" and had made that intersection a natural place to stand.[8]

During his lifetime, Lashley delighted in creating myths about himself, rewrit-ing his own history and obscuring the steps by which he reached his conclusions. According to the most tenacious of these myths, Lashley was a scientist who transcended time and place, a man without context. His science, he insisted, was motivated by no worldly concerns, but by a desire simply to understand the facts of brain function, mind, and behavior. He was, in the words of one of his admirers, a "scientist's scientist" who not only upheld but seemed actually to fulfill an ideal of purity, who pursued his research without an interest in its possible medical applications, in its social or moral value, or in its political meanings. Ultimately, he hoped, his science would be cleansed even of theoretical presuppositions.

The retrospective glorifications by the neuroscientists and by the historian Gardner, while they certainly reinforce Lashley's importance, do little to dispel this persistent myth of the man without context. Both cognitive science and neuroscience were late twentieth-century phenomena: the former began to gain momentum only by the mid-1950s; the latter was not officially founded until 1961. Neither took root and developed during Lashley's lifetime. By naming Lashley as their scientific forefather, the authors of these accounts achieve a curious feat: they place Lashley in the context of a future he could not have imagined and for the most part did not see. They perpetuate the image of the transcendent scientist by showing the great extent to which Lashley was ahead of his time, by excising him from his own early twentieth-century context and placing him in a later one. Neither of these accounts asks, or answers, the deeper

[7] Howard Gardner, *The Mind's New Science: A History of the Cognitive Revolution* (New York: Basic Books, 1985), 264.
[8] Eliot Stellar, "Physiological Psychology: A Crossroad in Neurobiology," in *The Neurosciences: Paths of Discovery*, eds. Frederic G. Worden, Judith P. Swazey, and George Adelman. (Cambridge: MIT Press, 1975), 363.

historical questions about Lashley and his work: why did he choose to do the work he did *when* he did, before it was vindicated by later approaches? Why did he decide to stand at the crossroads *before* the traffic had begun to roar?

More generally, this study asks: to what scientific and social contexts does Lashley belong? How did these contexts shape his approach to and solution of a seemingly transcendent scientific problem? To what, or to whom, in his own time and place, was he responding?

Born in 1890 and embarked on a scientific career by 1910, Karl Lashley was the heir to two different nineteenth-century scientific traditions: a neurological tradition, based in the medical clinic and devoted to the localization of mental functions in the brain; and a neurophysiological tradition, based largely in the laboratory and dedicated to the experimental study of the reflex. Lashley accepted neither of these traditions wholesale; rather, he borrowed and combined key insights from both in creating his own science. From neurology, he took the emphasis on the brain and its functions, but removed it from its clinical setting. From neurophysiology, he borrowed the centrality of the laboratory, but jettisoned the reflex concept. With this new combination of old scientific tools, Lashley set out to reform the field of psychology, to turn it from a semi-humanistic discipline into a rigorous, decisively materialistic, laboratory-centered experimental science of brain and behavior.

Of the two traditions, Lashley was more closely associated with neurology. In the nineteenth century, this science was dominated by the belief that the various mental functions were located in specific regions of the cerebral cortex; thus the tradition was often called the science of "cerebral localization." In its most familiar and often caricatured form, localization was represented by phrenology, Franz Josef Gall's attempt to find the seat in the brain of such complex mental functions as memory, reason, intellect, and will. Gall derived his conception from close anatomical, behavioral, and physiognomical studies on his fellow human beings, as well as from comparative anatomy. His idea was rapidly popularized by his followers as the notion that one's strengths and weaknesses could be divined by "reading" the bumps on one's skull.

By 1820, phrenology had begun to fall into disrepute for a variety of reasons, among them the 1824 discovery by the French physiologist Pierre Flourens that the brain actually functioned as a whole. Pursuing a unitary conception of mind, Flourens argued that the brain too must be unitary, not a mosaic of discrete capacities (an argument often considered an anticipation of Lashley's a century later).

By the 1860s, however, the localizationist theory had returned to fashion: Gall's doctrine of separable mental faculties was revived with the discovery by a French neurologist, Paul Broca, that the advanced mental faculty of "articulate language" could be precisely located in the third left frontal convolution of the cerebral cortex. Broca made this discovery through a series of famous studies of patients with aphasia, or inability to speak, apparently brought on by damage to the center in the brain controlling expression of language.[9]

Meanwhile, other discoveries in nervous function, coming out of the neurophysiological tradition, were taking place. The physiologists Charles Bell in 1811 and François Magendie in 1822 independently discovered that sensory and motor functions could be assigned precisely to different areas of the nervous system: to the dorsal and ventral roots, respectively, of the spinal nerves. Thus the dorsal roots were responsible for identifying and classifying incoming sensations, while the ventral roots were the source of the motor responses to those sensations. In the 1830s, the work of Johannes Müller and Marshall Hall demonstrated that this connection between sensation and motion, between stimulus and response, was the actual physical basis underlying reflex behavior. The physical connection between the two types of nerves was the actual physical path along which a reflex traveled; Müller and Hall believed that these reflex arcs, coursing through the nervous system, connecting sensation with motion, were at the heart of all automatic but purposive behavior.[10]

In the latter part of the nineteenth century, the reflex concept became, in the hands of neurologists and neurophysiologists, an immensely powerful tool to sort through the complexities of nervous function. The challenge was to determine the extent of the usefulness of the reflex concept. It worked perfectly well to explain automatic or lower level behaviors, but could it be used in explaining higher mental functions as well? Everyone agreed that reflex arcs coursed through the subcortical nervous system, but how high up the spinal cord could the concept be applied? Was the cerebral cortex itself implicated in the reflex arcs? Did reflexes course through the brain? Were reflexes at the basis of mind?

A surprising number of neurophysiologists and neurologists from a diversity of national and cultural backgrounds answered this question in the affirmative, and were supported by experimental evidence. In one of the earliest expositions of the idea, the Russian physiologist Sechenov, in his 1863 work "Reflexes of the Brain," established the reflex as "the basic unit of *all* psychological functions."[11] In the 1870s, the British neurologist John Hughlings Jackson, following his teacher Thomas Laycock, as well as the British philosopher-psychologists Alexander Bain, Herbert Spencer, and G. H. Lewes, also began to turn the reflex concept to an explanation of cerebral function. The brain, Jackson theorized, was itself a sensory/motor apparatus that operated through the creation of reflex con-

---

[9] For a history of brain localization, see Anne Harrington, *Medicine, Mind and the Double Brain* (Princeton: Princeton University Press, 1987), and Robert M. Young, *Mind, Brain and Adaptation in the Nineteenth Century* (New York: Oxford University Press, 1970).

[10] For a history of the reflex concept, see Roger Smith, *Inhibition: History and Meaning in the Sciences of Mind and Brain* (Berkeley: University of California Press, 1992), especially chapter 3; and Franklin Fearing's 1930 classic, *Reflex Action: A Study in the History of Physiological Psychology* (Baltimore: Williams and Wilkins).

[11] Smith, 108. Emphasis in original.

nections. No ideas were innate to it; rather, complex ideas developed within it through connections, or associations, between sensations and the motor responses they produced. Jackson's belief was bolstered by an 1870 discovery by two German neurologists, Fritsch and Hitzig, that motor responses could be obtained through the electrical stimulation of certain areas of the cortex; thus there were "motor centers" in the brain. In 1873 there was further confirmation by the British neurologist David Ferrier, who, using Fritsch and Hitzig's method, discovered sensory centers in the cortex as well.[12]

Despite these successes, by the mid 1870s John Hughlings Jackson had become aware of a serious problem with the reflex conception. How was it possible to understand or explain the transition from a "lower" function, such as a reflex arc linking sensation and motion, to a "higher" mental faculty, such as the language center that Broca had identified? Jackson was faced by a gap between physiology and intellect, a gap that he ultimately decided he was unable or unwilling to bridge. He concluded that, through the laws of association, the processes of sensation and motion "somehow" produced higher level ideas. To get around having to nail down this "somehow" any more specifically, Jackson adopted a philosophy called "parallelism" that allowed his neurological work on the brain to proceed without miring it in questions about how exactly mind emerged from the nervous system. Intellect, Jackson maintained, could not really be the subject of scientific inquiry; one must operate under the assumption that it was somehow produced *"while"* the various physiological processes were occurring in the brain and nervous system. While sensation and motion were becoming associated, while reflexes were forming, simultaneously, in a parallel mental world, ideas were forming in the brain. Jackson wrote:

> I do not trouble myself about the mode of connection between mind and matter. It is enough to assume a parallelism. That along with excitation or discharges of nervous arrangements in the cerebrum, mental states occur, I of course admit; but how this is I do not inquire; indeed, so far as clinical medicine is concerned, I do not care.[13]

In 1874, Carl Wernicke, a German neurologist and student of Theodore Meynert, set out to solve the problem of Jackson's elusive "somehow": to bridge the gap between Broca's advanced faculty of articulate language and Jackson's lower sensory/motor processes. Again, he posed the question: how, exactly, did the laws of sensation and motion produce the higher powers of intellect? Following Meynert and Jackson, Wernicke accepted the idea that the brain operated on reflex principles; what was disrupted in aphasia, for example, was the "psychic

---

[12] Harrington, 206–210, and Young, 204–210.
[13] Quoted in Young, 208.

reflex arc necessary for the normal speech process."[14] All mental functions were made possible through such reflex connections. Then, addressing himself specifically to the problem of the gap, Wernicke theorized that what could be localized in the brain were not the complex mental functions, like "memory," "intellect," or various character traits, but much simpler sensory and motor "traces" that he envisioned as "primitive memories." These traces could be found in those brain areas directly contiguous to the sensory and motor areas. When the traces were linked together by reflex arcs, the "higher" mental functions, such as the ability to speak articulately, emerged. During the remainder of the decade, investigators building on Wernicke's conception localized a wide array of mental functions in the cortex, finding centers for reading, writing and calculation, and associating the loss of each center with a corresponding clinical deficit.[15]

Despite Wernicke's modifications to the Broca/Jackson problem, and the success he had in gaining a scientific following, he had managed only to narrow the gap, not to close it. He had still not specified how the complex mental functions emerged from associations among the primitive memory centers. Anne Harrington notes the vagueness built into Wernicke's conception: "These primitive memories . . . served as the basic units that, interacting via the fiber tracts according to the laws of 'association', gave rise (somehow) to the rich complexity of mind."[16] Wernicke had redrawn the line between physiology and intellect, but still could not explain how the one became the other. In Harrington's words, "The association thus created resulted in the birth of an 'idea' – how is not really clear – that in turn, was associated with other ideas giving rise to proper sequential thought."[17]

Thus the gap between lower functions and higher remained an inescapable feature of the reflex theory of brain and mind. As long as complex mental functions were believed to be reducible to simple elements, there would be the problem of figuring out how the simple became transformed into the complex. None of the nineteenth-century localizationists managed to solve this problem, loyal as they were to the reflex concept. How to solve or get around the problem, then, became the major challenge facing twentieth-century American investigators in the mind-brain sciences.

When Lashley took on the problem in 1920, he identified the reflex concept, which the localizationists had borrowed from neurophysiology, as the source of the difficulty. His first major break from localizationist tradition was to reject this ubiquitous and powerful concept. In doing so, he rejected the disjunction between the "lower" sensory and motor functions, with their well understood basis in biology, and the higher mental powers, which "somehow" emerged from connec-

---

[14] Quoted in Harrington, 72.
[15] Ibid., 73.
[16] Ibid., 72.
[17] Ibid., 73.

tions or associations among lower functions. Mind, Lashley argued, did not get built up "somehow" from reflex connections. The Broca/Jackson/Wernicke gap was the artificial outcome of a false theory; their reliance on the elusive "somehow" was a result of their antiscientific approach to mind. The higher functions need no longer be approached only by way of the lower, but could be analyzed directly by experimental investigation. They need no longer exist vaguely on the other side of an unbridgeable gap.

Second, more positively, Lashley replaced the diverse mental faculties identified by the localizationists with a singular mental quality: intelligence. Crucial to the success of his program was his construction of this term. Unlike the localizationists, Lashley did not envision intelligence as an immaterial quality that emerged from physical connections – a notion that relied on some indefinable turning of the hinge from the physical to the mental, from the physiological to the psychological. Instead, from the start Lashley defined intelligence as a mental and a biological entity that had both physiological and psychological dimensions. It could be represented not only in the behavior of an animal or a person, or in their performance on tests, but also by the physical amount of functional brain mass that they possessed. Thus intelligence was for Lashley a term that did crucial work in bringing together the mental and the physical, mind and body, psychology and neurology – realms that the nineteenth-century localizationists, whatever their intentions, had ultimately kept separate. Intelligence was a term that functioned equally well, and that was intended to function well, in all of these realms; it helped Lashley to demonstrate the fundamental impossibility of holding spirit separate from matter. Moreover, in choosing the term, Lashley assigned a specific technical meaning to, and devised specific scientific measures for, a word that was already familiar to the layperson, that already had conventional social meanings and connotations.

With the rejection of the reflex concept and the reification of intelligence, Lashley fundamentally revised the nineteenth-century parameters of the mind-body problem. Parallelism, the belief in the separateness of the physical and the mental, the ultimate recourse of nineteenth-century neurologists and neurophysiologists alike, was simply not an acceptable answer for him. It was a cure worse than the disease: not a solution to the problem, but the admission of a deep and abiding ignorance. The only possible answer, Lashley believed, was a thoroughly materialistic one: everything in the world, including mind, was essentially the interactions of matter. There was no room in his science for an immaterial intellect, or even for the idea that thought somehow "emerged" from the brain.

Though he parted ways conceptually from the nineteenth-century neurologists and neurophysiologists, methodologically Lashley remained much in their debt. The method that he spent almost four decades developing and perfecting was an eclectic mixture of the two nineteenth-century traditions. From his teacher Shepherd Ivory Franz (who was himself a student of the localizationists), Lashley borrowed the method of cortical ablation, in which different areas of the cerebral cortex are systematically destroyed and the effect on behavior assessed. But while the localizationists, being mainly clinicians, used human beings as their subjects and therefore had to wait until nature or some misfortune produced someone with a damaged brain, Lashley, now following the neurophysiologists, used animals – almost exclusively rats – as subjects, and thus could control the area and extent of brain lesions himself. Like the Russian physiologists Sechenov and Pavlov (whose influence in particular came to him through his teacher John B. Watson), Lashley removed the site of his science from the medical clinic and located it firmly in the laboratory.

With these methodological resources – the use of cerebral ablation, the emphasis on animals, the centrality of the laboratory – held in place by a thoroughgoing materialism, Lashley built a science he called neuropsychology.[18] Its purpose was to reveal the relationship between brain function and behavior, particularly intelligent behavior, in rats, though its results could be extrapolated to higher species, including humans. The method on which it rested was simple in conception, though over the course of a forty-year career, Lashley elaborated it to tremendous proportions. First, he ablated a portion of the rat's cerebral cortex. Then, after giving the rat a chance to recover from the operation, he tested its ability to learn or remember a series of tasks, such as running a maze, distinguishing between two patterns, or tripping the lever on a specially designed "problem box."

Lashley found, much to his surprise, that the rats, despite missing parts of their brains, did not lose their ability to function intelligently, and that in fact up to 50 percent of the cortex could be extirpated before learning and memory became severely impaired. He concluded from his experiments that function was not localized in discrete portions of the brain, that ideas and memories were not stored in single cells like jewels in individual jewel cases; rather, function was distributed throughout the brain. Lashley summarized his findings in two principles that have become synonymous with his name: equipotentiality, that all parts of the brain are capable of carrying out all functions; and mass action, that only a certain amount of brain is necessary to ensure normal function.

Though Lashley derived the methods and strategies of his own science mainly from traditions in neurology and neurophysiology, he never associated himself professionally with those disciplines. Rather, his professional allegiance was al-

---

[18] Today the term "neuropsychology" signifies a branch of psychiatry that locates the source of mental disorders in brain lesions. While the mind-brain correlation clearly owes something to Lashley's science, the clinical application was utterly foreign to his perspective. To understand the meaning of Lashley's neuropsychology, we must cleanse the term of its modern connotations. For Lashley, neuropsychology was intended to illuminate the functions of the *normal* brain. For a history of the term and Lashley's use of it, see Darryl Bruce, "On the Origin of the Term 'Neuropsychology,'" *Neuropsychologia* 23.6 (1985): 813–14.

ways to psychology: his closest associates were psychologists; he held positions in psychology departments; the professional organizations he joined and led were psychological ones. Yet he wrote again and again that the science of psychology needed reform; that destructive and antiscientific tendencies existed among psychologists; and that only neurology and neurophysiology could combat those tendencies. Consequently, throughout his career, Lashley cast himself in an oppositional role with respect to the psychological community of which he was a professional member, resisting every new trend in psychology with the tools he had borrowed from the brain sciences, arguing that only a close relationship with biology could rescue psychologists from intellectual bankruptcy.

In the first half of the twentieth century, American psychology was a field in search of legitimacy: in search of a subject matter, a method, and the right to call itself a scientific discipline. It was, in the words of William James, the study of human mental life; but how that life was to be examined, in what it consisted, and who had the authority to pronounce upon it, were all unsettled questions. One answer issued from the laboratory of Wilhelm Wundt in Leipzig, and was transplanted to America by E. B. Titchener and his disciple Edwin G. Boring. Their approach relied on the method of introspection, by which a human subject, often the psychologist himself, reported on the workings of his own mind, in response to a series of controlled stimuli. Introspection was therefore a process that took place in a laboratory and that required training in distinguishing the responses worthy of report.

James scoffed at this method, dismissing it as "brass instrument psychology" (in reference to the laboratory apparatus that created the stimuli) and arguing that introspection was by definition impossible. In a memorable passage, James wrote:

> Let anyone try to cut a thought across in the middle and get a look at its section, and he will see how difficult the introspective observation of the [thought processes] really is. The rush of the thought is so headlong that it almost always brings us up at the conclusion before we can arrest it. Or if our purpose is nimble enough and we do arrest it, it ceases forthwith to be itself. . . . [as] a snowflake crystal caught in the warm hand is no longer a crystal but a drop . . . . The attempt at introspective analysis . . . is in fact like seizing a spinning top to catch its motion, or trying to turn up the gas quickly enough to see how the darkness looks.[19]

Instead, James, Peirce, Dewey and other American philosopher-psychologists who called themselves "pragmatists" believed that consciousness should be considered not a static thing whose structure needed description, but an ongoing process, a function, whose purpose was the progressively greater adaptation of the organism to its environment. James believed that such a psychology could not

[19] William James, "On Some Omissions of Introspective Psychology," *Mind* 9, no. 33 (January 1884): 3.

properly be pursued in a laboratory, but instead had to come to terms with the lived experience of real people, with their emotions, thoughts, and beliefs.

Whether these different theoretical perspectives actually had an impact on the practice of psychology is not clear. Whatever theoretical "school" they belonged to, most psychologists in this human-centered science used both introspection and close behavioral observation of other human beings to understand the structure and function of the "mind."[20] For the most part, psychologists did experimental work on the human sensory apparatus, either the experimenter's own or another person's; one of the most common types of study, for example, was the testing of a person's reaction time to a stimulus.

It would be difficult to exaggerate the epistemological leap required to go from this psychology, focused on the depths of one's own mind or the capacities of another's, to a science that used animals as its sole objects of study. In the early twentieth century, psychologists who studied animals were in a distinct minority. Most of psychology's time-honored methods and assumptions were inapplicable to the animal domain. Animals could not give introspective reports; whether they possessed consciousness in the human sense was a matter of debate. However, some of the other psychological methods – the reaction-time studies, for example – could be adapted to animal subjects. Consequently, in the early decades of the twentieth century, a group of comparative psychologists answered the challenge to create a "science of the animal mind."

They began by arguing that all speculations about the existence of the animal mind must be banished from the purview of their science. The focus of a truly scientific psychology must instead be on outwardly observable behavior. While George John Romanes, an early comparative psychologist and follower of Darwin, had tried to imagine how animals must feel and think, on a human model, his intellectual descendants (C. Lloyd Morgan, Edward L. Thorndike, and especially John B. Watson) were interested only in the external manifestations of these internal mental processes. What these internal processes were, indeed, whether they existed at all, made absolutely no difference to the study of behavior. Mentation was not an objectively observable phenomenon, and so had to be dismissed from scientific study. At their most radical, the "behaviorists" (as Watson called them) believed that their new psychology could apply as well to human beings as to animals. Reorienting the methods and purpose of psychology, the behaviorists brought comparative psychology out of its stagnant backwater and put it at the forefront of psychology as a whole.[21]

Lashley, however, was never able to find a comfortable home in psychology, whatever its orientation. As an animal psychologist among the human-centered

---

[20] In addition to introspection, other important approaches in psychology included G. Stanley Hall's developmental psychology and Floyd Allport's social psychology.

[21] See John M. O'Donnell, *The Origins of Behaviorism: American Psychology, 1870–1920* (New York University Press, 1985).

psychologists, he was marginalized. Yet he was equally ill at ease among the behaviorist comparative psychologists. The behaviorists held that all internal processes were beyond the reach of their science; both the mind and the brain of the organism they treated as a black box, their only concern being its inputs and outputs, not what was going on inside. Lashley's focus on the brain and on the internal, biological correlates of behavior, his desire to bring about a coalescence of psychology and neurology, went directly against the grain of behaviorism. The behaviorists were concerned to defend the autonomy of their science, its independence from older and more securely established disciplines. They had, they believed, broken psychology's ties with philosophy by ridding it of such inexact concepts as "consciousness" and "mind." Having established its scientific credentials, they were now equally wary of enslaving it to other disciplines: biology, physiology, neurology. Lashley's emphasis on the necessarily close relationship between psychology and biology, then, ran directly counter to the search for autonomy among his fellow comparative psychologists.

Above all, behaviorism was supposed to transform psychology into a useful science, a science aimed at the prediction and practical control of behavior. Condemning the introspectionists' solipsistic inward gaze and their predecessors' fruitless ruminations on the "animal mind," the behaviorists maintained that though psychology was an experimental science, it should not limit itself to the narrow confines of the laboratory. Echoing James and Dewey, the behaviorists argued that psychology must take real life as its subject; that its practitioners must get out of the laboratory and onto the street, that they must start observing how people really act.

Watson was an eloquent advocate of the ideal of a useful, relevant, socially applicable science; but that ideal was taken up by many psychologists who did not necessarily agree with Watson's specific formulations. The rise of the mental testing movement in the 1910s was a prime example of psychologists' interest in a practical science that would yield practical results. Inspired by their success in organizing a national campaign to determine the intelligence of army recruits, psychologists began to envision themselves as expert consultants on the problems of society. A society at peace needed them quite as much, they argued, as did a society at war. Mental tests could be administered to the general population and, with psychologists on hand to interpret the results, people could be guided to the stations in life for which they were best suited. Once they were there, psychologists could advise them on how to live more productive lives. For Robert M. Yerkes and other believers in psychology as a practical science, the truly professional psychologist was the one who turned his knowledge toward applications.[22] Applied and consulting psychologists appeared in advertising firms, on the boards

---

[22] See Robert M. Yerkes, "Man Power and Military Effectiveness: The Case for Human Engineering," *Journal of Consulting Psychology* 5 (1941): 205–209.

of industry, in women's magazines. The rise of psychological expertise and its impact on all sectors and aspects of twentieth-century American society are among the most important themes in the history of the discipline.[23]

In response to the growing social role psychologists had begun to play, Edwin G. Boring – who had by the 1920s appointed himself the science's historian – tried to engineer a return of psychology to the laboratory, a return to the pure science that Wundt and Titchener had practiced. In his classic 1929 textbook, *A History of Experimental Psychology*, Boring declared that psychology in its most perfect form was an experimental, laboratory science, free of the need for social relevance, free of the context that shackled the less well-developed disciplines.[24] But in his reaction Boring did not mean to return to the introspective psychology of his teachers; rather, he chose as the new exemplar of this transcendent science Lashley's neuropsychology. For Boring, Lashley's work represented the highest form of experimentalism. Lashley, for his part, did his best to justify this belief, by resisting the trend toward applications and maintaining a standard of purity that denied the relevance of science to the problems of society.

But the double identity of psychology as pure and applied science proved difficult to house under one disciplinary roof; as a result, the first half of the twentieth century marked a key point of divergence. As the majority of American psychologists took the path toward social applications, they parted ways crucially with Lashley and his neuropsychology. As they defined psychology as the study of the individual's relations within society, their interest in Lashley's kind of science waned; the need for a coalescence of psychology and neurology mattered less and less. If the mind-brain problem went unsolved, few repercussions would be felt in the world of applied and consulting psychology. Applications could work without any understanding of or attention to the biological basis of behavior.

Gradually but irrevocably, those scientists who remained interested in unravelling the mind-brain connection ceased being psychologists. Like Lashley, they had to find a new name for themselves; ultimately they broke off and created their own new communities. They began to call themselves neuropsychologists; later, they became neuroscientists and cognitive scientists. This divergence between the society-centered science of psychology and the biologically based science of the mind/brain, the reasons for it and the effects of it, are recurring themes in this study.

Lashley's decision to pursue a pure, experimental, biological psychology, and his consequent position outside the dominant trends in psychology, was not a choice he ever regretted. In fact, he relished his outsider's stance. While being championed by Boring and holding positions of power and prestige within the

---

[23] See Ellen Herman, *The Romance of American Psychology: Political Culture in the Age of Experts* (Berkeley: University of California Press, 1995).

[24] E. G. Boring, *A History of Experimental Psychology* (New York: Appleton, 1929).

psychological community, Lashley nevertheless always managed to be the quintessential outsider. His opposition to the strong trend toward social applications was matched by his adamant refusal to uphold any of the variety of theoretical perspectives available in psychology during his lifetime. Though he worked as a research assistant to John B. Watson and became Watson's close friend, Lashley never considered himself a behaviorist; in fact, he spent the better part of his career arguing against the behaviorist tenets. Likewise, for many years Lashley worked at the University of Chicago, where he came into close contact with the anatomist Charles Judson Herrick and the embryologist Charles Manning Child. Herrick and Child were leading members of the American school of psychobiology, a school of thought informed by the pragmatism and evolutionary functionalism of James and Dewey. Yet Lashley never joined their crusade to reform science and society. Though the Gestalt theorists, especially Wolfgang Köhler, also helped to shape his views, Lashley could never fully ally himself with them either. In the same way, Lashley was a vehement anti-Freudian throughout his career, even after the Second World War, when Freudian theory became popular among psychologists.

Lashley's outsider, oppositional stance was an attitude he cultivated assiduously, and it worked distinctly to his advantage. His anti-applications, atheoretical standpoint helped him maintain his own image as a neutral scientist and the image of neuropsychology as a pure, fact-driven science. Lashley deliberately rid himself of context and stripped his science of theory and applications in order to emphasize its status as an unblemished mirror of nature. This study will look beyond Lashley's oppositional stance, to understand why he opposed the perspectives in psychology that he did, and to explore the contexts, both scientific and social, that he did not oppose: to determine, that is, where he belongs. To see Lashley as the descendant of the localizationist tradition, its lonely defender in the hostile world of early twentieth-century American psychology, only to be vindicated later by cognitive science and neuroscience, is to construe the meaning of his neuropsychology far too narrowly, to miss much of the richness, complexity, and broader significance of his work. To conclude that the "equipotential brain" was his major contribution is to ignore the more radical implications of that idea. Lashley's emphasis on whole-brain functioning was only the means to an end, that end being the advancement of a scientific theory about intelligence, specifically about the hereditary nature of intelligence.

Lashley's answer to the mind-body problem, his revision of the nineteenth-century parameters of the debate, was his definition of intelligence as a biological property. As his career progressed, he came increasingly to identify "biology" with heredity and, ultimately, with "genetics." The equipotential brain served to support the argument: because the brain functioned as a whole, because all parts of it were in constant activity, whatever level of brain function one was granted at birth, one had for life. There was no way to improve on the brain's function; there were no inactive portions waiting to leap into motion; mental capacity was unchangeable. Over the course of his career, Lashley's hereditarianism became an ever-present feature of his scientific work. It drove him to explore the essential similarity between learned behavior and instinct, at a time when the concept of "instinct" had been dismissed by most psychologists. It motivated him to forge hitherto unnoticed connections with hereditarian IQ testers, and to give their work a respectable grounding in biology. His hereditarianism inspired his study of the biological basis of sexuality. And it reinforced and was reinforced by the profoundly conservative and nativist vision of American society that he shared with many of his students and followers.

Placing Lashley's neuropsychology in its scientific and social contexts, far from "debunking" his science or exposing his scientific "feet of clay," reveals that Lashley was centrally involved in a broad conversation about mind and behavior: a conversation not simply about brain function in rats, but also about human intelligence, its hereditary basis and possible alterability, and about the proper organization of society. This conversation involved not only psychologists and other social scientists, but also biologists in many different areas (including neurology, physiology, embryology, and genetics) throughout the twentieth century; in different forms, it has persisted to the present day. Exploring Lashley's context only deepens his relevance for us, who at the turn of the twenty-first century are still grappling with our own variations on the questions that he was asking. How is hereditary endowment related to behavior? How do brain structure and function mediate between genes on the one hand, and intelligence or mental capacity on the other? Lashley's project – to excavate the "natural," indwelling essence of the socialized human being – has hardly been forgotten: today it is more popular than ever, reappearing in such guises as the "gay gene" and *The Bell Curve,* and in such new interdisciplinary sciences as behavioral genetics, sociobiology, and evolutionary psychology. It is therefore more important now than ever to understand the assumptions that guided Lashley's project and those that his project in turn supported, and to reconstruct the conversations of which he was a part.

An interpretation that locates Lashley's neuropsychology at the center of hereditarian conversations about mind, behavior, and human society requires that the history of twentieth-century psychology be considered together with the history of biology. Only by exploring the deep connections between the histories of these two sciences can Lashley be seen to "belong"; only then can we see the context into which he fits; only then can we reveal not only the important trends that he opposed, but also those, equally important, that he supported. Only then does his place in early twentieth-century American science make sense. But such interpretations that cross disciplines have been largely missing from the history of science, especially from the history of the life and human sciences, which tend to focus on professionalization and the rise of disciplines. Lashley does not fit into such a story; but that does not mean that he does not fit anywhere at all. Lashley's neuropsychology belongs, rather, to the history of the space between disciplines. By telling the kind of story that not only accommodates Lashley but also makes his contribution central, this study provides a new perspective on the histories of biology and psychology: it tells a new story about the American development of the sciences of mind, brain, and behavior in the first half of the twentieth century.

1

Lashley and Jennings:
The Origins of a Hereditarian

Prologue

In an obituary of his mentor, Frank Beach described the life and career of Karl Spencer Lashley as the embodiment of many contradictions. Lashley, Beach wrote, was a "[f]amous theorist who specialized in disproving theories, especially his own," and an "[i]nspiring teacher who described all teaching as useless."[1] Perhaps the most astonishing of these ironies is that Lashley was an "[e]minent psychologist with no earned degree in psychology." Indeed, Lashley received formal training neither in neurology nor in psychology, the sciences which became the foci of his mature work. Lashley himself noted:

> My training has been atypical for psychologists. As an undergraduate I specialized in comparative histology; my master's thesis was in bacteriology and my doctor's in genetics. . . . I did not choose psychology as a career until two years after the Ph.D. I never attended a course in physiology or neurology, which have become my major interests.[2]

For Beach and the historians who have followed him, the problem has been to understand adequately how Lashley became renowned in fields for which he had little preparation. Most recently, in fact, Darryl Bruce has argued that Lashley's "shift" from his undergraduate work to his mature research occurred in several

---

[1] Frank A. Beach, "Karl Spencer Lashley, June 7, 1890–August 7, 1958," *Biographical Memoirs of the National Academy of Sciences* 35 (1961): 162.

[2] Karl S. Lashley in response to Lauren G. Wispe's questionnaire, "Factors in Psychological Leadership," undated, Lauren G. Wispe Papers, Box M699, Archives of the History of American Psychology, Bierce Library, University of Akron, Akron, Ohio. By "genetics" Lashley meant the study of inheritance – specifically, of inherited patterns of behavior in protozoa – and the demarcation of inherited patterns from those that were environmentally influenced. This was the focus of his work with Jennings. Between 1900 and 1915, the chromosomal theory of inheritance was being established by Thomas Hunt Morgan and others at Columbia University, according to which traits inherited in Mendelian fashion had their source in genes located on chromosomes. Thus the Mendelian "factors" responsible for different hereditary traits were given a material basis in living cells. Though Lashley, as Jennings's student, doubtless knew of Morgan's work, his own research was not in the physical processes by which traits were inherited, not in "genetics" in that sense, but rather in the patterns according to which different traits were passed from parent to offspring. Thus Lashley's graduate work with Jennings might more properly be called research in heredity, or hereditary traits and behaviors, to distinguish it from the kind of work Morgan was doing.

distinct steps, from bacteriology, to zoology and genetics, to comparative psychology, to learning, and finally to the neural basis of learning.[3] In Bruce's portrayal, Lashley was influenced at crucial points by Herbert Spencer Jennings, John B. Watson, and Shepherd Ivory Franz.

At the same time, however, there is a competing image of Lashley's life that underscores not its shifts and contradictions, but rather its continuities. In the introduction to their collection of Lashley's papers, Beach and his colleagues designate the earliest period of Lashley's career as "laying the foundations" for what came later.[4] Bruce echoes this description in his article. In his own accounts of his career, Lashley stressed the continuity of his interests: he said that during his freshman year at the University of West Virginia, he found his "life's work" in a course with the neurologist John Black Johnston. When Johnston was replaced by Albert M. Reese, Lashley worked as Reese's assistant and proposed to Reese that he try to trace all the connections between the neurons of the frog's brain. He said later, "I think almost ever since I have been trying to trace those connections."[5]

Doubtless there is truth in both of these images. Despite the fact that Lashley did not actually chose psychology as a career until two years after he finished his Ph.D., he had an abiding interest in the problems that occupied him in his maturity. Here I will take a fresh look at Lashley's early life and education, the beginnings of his interest in the problem of mind, brain, and behavior, and the roots of his hereditarianism, and will suggest some hitherto unseen continuities between his early career and his mature investigations.

Lashley's Early Life

Lashley was born in 1890 in Davis, West Virginia, a small country town where his father ran the family's general store, managed a bank, and served as mayor and postmaster.[6] His mother, a teacher before her marriage, spent her considerable intellect and energy amassing a library, instructing the local ladies in a variety of subjects, and being the best friend of her only child. As a boy Lashley was a loner, preferring the company of books and pets to that of other children. During the 1890s the family moved several times but landed finally in Seattle, whence, struck by gold fever, they went north to Alaska to prospect.[7]

---

[3] Bruce, "Lashley's Shift," 27–45.

[4] Frank A. Beach, Donald O. Hebb, Clifford T. Morgan, and Henry W. Nissen, eds., *The Neuropsychology of Lashley: Selected Papers of K. S. Lashley* (New York: McGraw Hill, 1960), vii.

[5] Lashley quoted in Beach, "Karl Spencer Lashley," 169.

[6] I have collected the details of Lashley's early life mainly from an interview that Anne Roe conducted with him in 1950 (Anne Roe Papers, Collection B:R621, American Philosophical Society Library, Philadelphia); also from the obituary by Frank A. Beach, "Karl Spencer Lashley," *Biographical Memoirs,* 162–204.

[7] Transcript of Lashley interview, page 1, Roe Papers, APS.

By 1898, when their luck in this venture had run dry, they returned to Davis. There, tutored by his mother in the evenings, Lashley began a period of intensive schooling. At the age of twelve, influenced by his reading of Robert Green Ingersoll and Thomas Paine, Lashley developed an active antipathy toward religion: on shooting expeditions with his friends he was given to pointing a gun in the air and proclaiming that he wished to put a bullet into God. In high school his favorite subject was Latin; but when he entered the University of West Virginia intending to major in it, an adviser persuaded him to switch to English instead.[8]

All of these plans fell by the wayside when Lashley by chance happened onto a course in zoology taught by the neurologist Johnston. Almost at once he changed his major to zoology, and noted with pride many years later that his freshman grades were "Zoology, 96, Military Science, 96, Latin, 75, English, 23."[9] Johnston's successor Reese employed Lashley as his laboratory assistant and encouraged the young man to be his own teacher, handing him a sheep brain and telling him to work up a course in comparative neurology. Reese also took his protege to hear a Unitarian pastor, but failed to shake his firm belief in materialism. Lashley took only one college course in "mental philosophy" (including its obligatory overview of neurology), noting later with disdain that it "[d]idn't even come up to James."[10]

Lashley wrote his undergraduate thesis on comparative histology, and received an A.B. in 1910. The following year he won a teaching fellowship in biology at the University of Pittsburgh. The department was "all bacteriology," and after toying with the idea of studying the effects of pollution on the local rivers, Lashley wrote a master's thesis on the bacteriology of rotting eggs. He took several courses in psychology, but they involved introspection, not the neuropsychology in which he eventually specialized. Later in life, he recalled that by this point he had already begun to formulate the mechanistic theory of consciousness that he elaborated in the early 1920s. In Pittsburgh he also began to manifest a sensitivity to race and became an ardent segregationist; years later he recalled having seated black and white students separately in his laboratory sections.[11]

The summer of 1911 Lashley spent at the Cold Spring Harbor Laboratory working on the genetics of *Stentor* under the very minimal supervision of Charles Davenport. There he associated mainly with geneticists, among them A. H. Sturtevant and Sewall Wright. With a forlorn hope that their son might become a doctor, Lashley's parents were determined that he should attend the best university in the area, and in the fall he registered in zoology at The Johns Hopkins University. He met Jennings and enrolled in his courses, and Jennings employed him as an assistant in his work on *Hydra* and *Paramecium*. Watson had also just arrived at

[8] Ibid., pages 2, 4–5.
[9] Ibid., page 6.
[10] Ibid., page 9.
[11] Ibid., page 9.

Hopkins and Jennings suggested to Lashley that he take psychology as a minor. This he did, as well as a second minor in psychiatry with Adolf Meyer.

Jennings and Lashley collaborated on two studies of the effects of conjugation on size and vitality of *Paramecium,* which were published in 1913 in the *Journal of Comparative Zoology.* Lashley's dissertation, also under Jennings's direction, was on inheritance in asexual reproduction in *Hydra;* this and a follow-up study appeared in the same journal, in 1915 and 1916 respectively. During these years Lashley also began a collaboration with Watson. The two soon formed a close friendship; Lashley helped Watson read the German psychologists, and Watson got Lashley drunk on highballs.[12] Together they went to the Dry Tortugas and worked on the homing and nesting behavior of noddy and sooty terns. Lashley received the Ph.D. in genetics in 1914 and a fellowship to remain at Hopkins for a year after that. He continued to work on *Hydra,* but became increasingly drawn into Watson's behavior research. For the following two years, until 1917, Jennings arranged a scholarship for Lashley to continue working in psychology. With Watson, Lashley made studies of conditioned reflexes; their work on the role of the larynx in thinking was used by Watson as the basis of his 1915 Presidential Address to the American Psychological Association.

It was at the meeting of this association that Lashley met Franz, a psychologist at Government Hospital for the Insane in Washington, D.C., and heard him speak on habit retention after frontal lobectomy. Lashley suggested that they collaborate on a study of rat behavior, in order to determine the amount of training required to reduce a habit to subcortical representation in the brain. Lashley proposed that he train the rats and that Franz perform the brain operations. Franz agreed, and Lashley took weekly trips to Washington with the rats; he later worked as Franz's assistant at Government Hospital for the Insane. Under Franz's direction, Lashley spent a year engaging in disparate researches:

> My first day was probably typical of the man and his methods. . . . he took me to a burlesque show and the next morning he took me to an autopsy and he let me help with the brain operation in the afternoon and then he took me over to help photograph the distribution of hair in fat women.[13]

By this time, Watson had moved his laboratory to Meyer's Phipps Clinic in the Hopkins Medical School, intending to take up the study of infant reflexes; Lashley began as his partner, but reported later, "I hated the damn brats from the start." Ultimately Watson found another assistant, and Lashley concentrated full time on the work with rats. When the First World War began, Lashley was excused from the army because of poor vision; in any case, he later reflected,

[12] Ibid., page 12.
[13] Ibid., page 14.

I had no desire to get into the army. I was violently pro-German. If I had had to do anything, I would have gone to Germany and enlisted in the German army. My whole setting was toward the Germans. I was interested in the German evolutionist[s] and I had read Nietzsche and so on and my sympathies were all on that side. If I could have shot a Frenchman, I would have done so gladly.[14]

In 1917 a job opened up at the University of Minnesota under Robert M. Yerkes. Hesitant about the opportunity, Lashley told Jennings that Yerkes had a reputation for being difficult to get along with. " 'You aren't so easy yourself,' " Jennings replied. Lashley took the appointment, but spent only a short unhappy time there. Yerkes stayed in Washington, D.C., to help establish the National Research Council, and never assumed his chairmanship; the department was poorly supervised, and Lashley couldn't adjust to his heavy teaching schedule. He resigned twice and was only prevented from returning to Baltimore by the dean, who happened to be his undergraduate teacher Johnston. In the summer of 1918 Lashley married Edith Ann Baker, a promising pianist, who was then stricken by a debilitating asthma on their honeymoon. Lashley's salary was barely enough to keep her in medication; and as he said later, he became desperate.[15]

Lashley got a reprieve from Minnesota when Watson enlisted his help on a project for the United States International Hygiene Board. Their task was to test the efficacy of films about venereal disease; traveling through small southern towns, Watson and Lashley advertised free showings of the films, distributed their questionnaires, collected them, and left town as rapidly as possible. It was Lashley's job to interview the townspeople a month later in order to gauge the effects of the films. This project lasted only a year, and by 1920 Lashley was back in Minnesota. This time, however, the new chairman of the department, Richard M. Elliott, took a personal interest in Lashley's research, provided him with the most modern facilities, and relieved him of much teaching responsibility. Lashley focused his investigations on the brain, and, finding his mature work at last, began one of the most fertile research periods of his entire career.

The Problem of Influence

The story of Lashley's life and work is the story of the rapprochement between psychology and "genetics," or – to put more clearly what "genetics" meant to Lashley – between psychology and hereditarianism. Throughout his career Lashley remained firmly convinced that the basis of behavior was hereditary and that intelligence was hereditarily determined. In an era during which psychology was dominated by behaviorism, he was a fierce opponent of the idea that behavior

[14] Ibid., page 16.
[15] Ibid., page 17.

could be conditioned by the environment. Later in his career he became sympathetic to Gestalt theory because of its stress on innate capacity, despite the fact that he also frequently denounced the Gestalt theorists as mystics.

Lashley's interest in inherited patterns of behavior has gone unexplored and unexplained by historians, and remains all the more mysterious because the major influence on Lashley is usually thought to be Watson, the self-styled founder of behaviorism. It was Watson, so the argument goes, who was responsible for interesting Lashley in psychology, who helped turn his attention from genetics to behavior. For example, in an edited volume of Lashley's papers, the first article included is a study of the nesting behavior of terns on which Lashley and Watson collaborated.[16] The implication is that Watson provided the starting point for Lashley's career in psychology.

But this leaves some important problems unresolved. Lashley never became a Watsonian behaviorist: soon after their collaboration began, Watson's and Lashley's views began to diverge, as Watson worked on conditioning reflexes and Lashley turned to the neurological basis of mind. By the mid 1920s, they were arguing with each other. How is this almost immediate and later complete divergence to be explained? How is it that Lashley, supposedly introduced to psychology by the founder of behaviorism, never became a true behaviorist? Why did Lashley remain throughout his career such an anomaly in psychology?

In order to answer these questions, Darryl Bruce points to Jennings as a sort of secondary influence on Lashley. Jennings shaped Lashley's general theoretical perspective, giving him not only "a violent aversion to vitalism," but an equally great distaste for oversimplified mechanistic formulations. While Watson was the major influence, then, responsible for the general trajectory of Lashley's career, Jennings was responsible for the rebellion from Watson. Watson introduced Lashley to the study of reflexes and animal behavior, while Jennings turned Lashley away from Watson's rigid stimulus-response principles and toward a more biological conception of behavior. This apportionment of influences is also reflected in the edited volume of Lashley's papers: Lashley's work with Jennings is represented by one brief selection that comes second, after the Watson/Lashley article. However, Lashley's work with Jennings actually predated his work with Watson and comprised several substantial research reports.

While it is certainly difficult, if not impossible, to determine the influence of one thinker on another, some greater exploration of the relationship between Jennings and Lashley is needed.[17] The portrayal of Jennings as a secondary influence on Lashley oversimplifies the role of Jennings, and obscures some deep similarities between Jennings's career and Lashley's. Not only do the connections

[16] Beach, Hebb, Morgan, and Nissen, eds., *The Neuropsychology of Lashley.*

[17] On the problems of determining influence, see Quentin Skinner, "Meaning and Understanding in the History of Ideas," *Meaning and Context: Quentin Skinner and His Critics,* ed. James Tully (Princeton: Princeton University Press, 1988), 45–7.

between Jennings and Lashley help to explain Lashley's unshakeable belief in hereditarianism, they also make the larger argument that the histories of biology and of psychology cannot continue to be written separately, as they traditionally have been. Jennings is conventionally considered to belong to the history of biology, while Lashley, if he belongs anywhere, fits best in the history of psychology. It has probably been in the interest of maintaining these distinctions that Watson has been accorded the major influence on Lashley: a psychologist, after all, would have been the one to influence his younger associate to spend his career in psychology.

But these biology/psychology distinctions must be examined more carefully. Jennings was at least as influential in turning Lashley toward psychology as Watson was. Before Lashley had even finished his Ph.D., Jennings had already done much to make the study of inheritance inseparable from the study of behavior. When Lashley arrived at Johns Hopkins, both Jennings and Watson encouraged him to study psychology, though they came to approach the subject from very different perspectives. A rebellious Watson did not turn Lashley away from genetics and from his mentor. On the contrary, both Jennings and Watson encouraged Lashley in the same direction, and Lashley's perspective always remained closer to Jennings's than to Watson's. Jennings, then, becomes the primary influence on Lashley at the beginning of his career, with Watson accorded secondary status. This reallocation of influence explains why Lashley opposed Watson and his brand of behaviorism, and became an anomalous hereditarian among psychologists.[18]

Lashley and Jennings

During the course of their careers, both Jennings and Lashley traveled between psychology and the study of inheritance. Lashley moved from inheritance in protozoa to research in mind and behavior of mammals. Jennings moved from research on the behavior of the "lower organisms" to the genetics of higher animals and human beings. He made the transition from animal behavior to

---

[18] Lashley's two mentors were not the only ones interested in comparative psychology during the 1910s. That they belonged to a much larger group with shared interests and, to some extent, common methodological commitments can be adduced from the advent of a number of new journals dedicated to the field. *The Journal of Animal Behavior,* under the editorship of Robert Yerkes, published its first issue in 1911. In addition to Watson and Jennings, its editorial board included such prominent comparative psychologists as Madison Bentley, Harvey Carr, Samuel J. Holmes, Edward L. Thorndike, Margaret F. Washburn, and William M. Wheeler. In 1917, the first issue of *Psychobiology* appeared, under the editorship of Knight Dunlap. Lashley published many of his early papers in these two journals. In 1921 the two journals combined to form the *Journal of Comparative Psychology.* Few of the psychologists who organized and published in these journals were "behaviorists" in the Watsonian sense; but the fact that they did organize these forums to share their work shows that the study of animal behavior was coming into its own as a legitimate branch of scientific psychology.

genetics in the 1910s, when he collaborated with Lashley on a series of studies of inheritance of form and behavior in protozoa.

In 1906, Jennings published *Behavior of the Lower Organisms,* a study of the reactions of various protozoa and metazoa to changes in their environment. He argued that all behavior was regulation, the attempt by the organism to relieve conditions that interfered with its normal internal processes. Through the "readier resolution of physiological states," organisms would "learn" the reactions that helped them to avoid noxious stimuli. Without having to run through their whole repertoire of reactions, they would immediately react in the way that was most adaptive.

In good Lamarckian fashion, these adaptive reactions would, with practice, became part of the organism's hereditary endowment. The organisms that learned the adaptive reactions most readily, that performed them immediately when presented with noxious stimuli, would survive to pass on this ability to their off-spring.[19] Thus, reactions that individuals had to learn through laborious trial and error, their offspring could perform naturally. In this way, learning becomes a factor in evolution; an individual's behavior could be responsible for the survival and progress of the species. Bringing psychology together with heredity, then, was an interest of Jennings's even before he began to work with Lashley.

Jennings opened *Behavior of the Lower Organisms* by defining behavior. This seemingly simple task became a central point of contention between Jennings and Watson, and later between Lashley and Clark Hull in the 1930s and 1940s. For Jennings, behavior was "the general bodily movements of organisms," both internal and external. "Behavior is a collective name for the most striking and evident of the activities performed by organisms."[20] This included, then, not only the progress of a *Paramecium* across a watch-glass, but also the processes of digestion that took place within its organelles. The definition became contentious, however, once comparative psychologists turned to the behavior of higher organisms. Was the action of a rat's nervous system part of its behavior? Or did behavior in that case include only the rat's outward activity? Were internal processes still "striking and evident" activities? And if not, where was the line to be drawn around those processes that constituted behavior?

Throughout his book, Jennings described the activities of the organisms with a naturalist's eye, in meticulous detail, and emphasized that their behavior had to be understood mechanistically. No purpose, final aim, or idea in the mind of the organism was necessary to account for its behavior, contrary to various vitalistic theories. The processes of regulation that comprised an organism's behavior were no different from those that took place in the inorganic world.[21]

[19] Herbert Spencer Jennings, *Behavior of the Lower Organisms* (New York: Columbia University Press, 1906), 325–7.
[20] Ibid., v.
[21] Ibid., 343.

Still, Jennings was not ready to rule consciousness out of court; he remained agnostic on the question of whether animals in general, and the lower organisms in particular, possessed consciousness. Sometimes it was useful in understanding their behavior to pretend that they were actually conscious. But there would never be any way to determine whether or not they actually were, just as one could never know whether other people have minds. For Watson and the behaviorists who followed him, this indeterminacy was enough to render the concept unscientific. Jennings, however, retained it because he believed that protozoan behavior was the prototype on which all behavior, even human, was modeled; as a result, the protozoa might possess some variant of human consciousness.

For Jennings, the behavior of the lower organisms was continuous with intelligent human behavior: the regulative activities displayed by protozoa were different from human action only in degree of complexity, not in kind. Jennings believed that the conclusions he drew from his studies of the lower organisms were directly applicable to the human case. No line separated the behavior of the lower organisms from that of other animals: the bases of consciousness and intelligence were located in protozoan behavior. The study of the behavior of protozoa, then, became a legitimate part of comparative psychology, and comparative psychology became the basis for all human psychology.

Jennings emphasized that the source of reactivity in the protozoa was internal: these animals did not react automatically to whatever stimuli were presented to them by the environment. Rather, they almost chose, in a way similar to that of higher organisms, whether and how to respond to the environment: the outward stimuli were mediated by the intrinsic spontaneity of the organism.

> The reactions produced in unicellular organisms by stimuli are not the direct physical or chemical effects of the agents acting upon them, but are indirect reactions, produced through the release of certain forces already present in the organism. In this respect the reactions are comparable with those of higher organisms.[22]

The organism was no stimulus-response machine; organisms did not respond in a stereotyped way to stimuli. For Jennings, the behavior of the protozoa was as spontaneous as that of humans; they were not buffeted about by the forces of their environment, but acted through the discharge of their own internal energy. Because protozoan reactions could not be stereotyped, they were never really predictable: they depended on the internal state of the organism at the moment of stimulation. "The reaction to a given stimulus depends on the physiological state of the organism . . . and physiological states are variable. This is true both for the infusoria and for man."[23] An organism's reactions to external agents were depen-

[22] Ibid., 261.
[23] Ibid., 280.

dent on its physiological states and on progressive changes in those states. "Whether a given change shall produce reaction or not, often depends on the completeness or incompleteness of the performance of the metabolic processes of the organism under existing conditions."[24] External factors never acted directly to change behavior, but they always acted by changing the physiological state, which then changed the behavior.

Moreover, in most cases the organism's own movement brought about the environmental change, so again the source of the behavior was internal. The intrinsic motivity of the organism determined what environments it would confront and how it would respond to those environments. "The organism *is* activity . . . . spontaneous activities of the organism . . . . are, perhaps, the most important factors in its behavior"[25] (emphasis in original).

Behavior of the Lower Organisms was Jennings's response to the work of Jacques Loeb, his main opponent in the study of animal behavior.[26] The issues that separated Loeb from Jennings continued to polarize the community of psychologists throughout the first half of the twentieth century. Did the organism respond mechanically to stimuli, in a predictable (and therefore controllable) manner? Did the same stimuli always elicit the same reactions? While Jennings argued for spontaneity and internal control, Loeb believed that environmental conditions produced the reactions of organisms. As Philip Pauly notes, Loeb and his defenders argued that Jennings's observations could never form the basis for a science; prediction and control were essential to Loeb's conception of science, as they were for Watson.[27] Indeed, after the publication of his book, Jennings's field, the description of the behavior of the lower organisms, languished, even though Jennings's colleague at Hopkins, Samuel O. Mast, continued working in it. Watson, meanwhile, carried Loeb's conception into the study of mammalian behavior.[28]

Jennings himself refocused his attention on genetics. This might seem a radical switch, but in fact Jennings's work in behavior had laid the groundwork for it. His emphasis on internal sources of motivity in the organism underscored constitution, rather than environment, as the determining factor in the behavior of protozoa. In a sense, then, his work in genetics was a deeper explanation of the factors comprising constitution. Although he eventually came to discuss the human species and its improvement through breeding, Jennings began his investigations into

[24] Ibid., 299.
[25] Ibid., 284–5.
[26] The famous debate between Jennings and Jacques Loeb has been analyzed by Philip J. Pauly in his biography of Loeb, *Controlling Life: Jacques Loeb and the Engineering Ideal in Biology* (New York: Oxford University Press, 1987). See also Pauly, "The Loeb-Jennings Debate and the Science of Animal Behavior," *Journal of the History of the Behavioral Sciences* 17 (1981): 504–15.
[27] Pauly, *Controlling Life,* 126–7.
[28] Ibid., 129, 172–7.

genetics in the 1910s by working on patterns of inheritance of form and behavior in protozoa.

During this decade, he was concerned to demonstrate that Darwinian evolution by natural selection actually occurred: that sexual reproduction produced a continuous array of variation, and that natural selection acting upon continuous variation could eventually lead to the production of different species. In 1913, Jennings collaborated with Lashley on the first of a series of studies to fulfill this aim. In "Biparental Inheritance and the Question of Sexuality in *Paramecium*," Jennings and Lashley questioned the theory that conjugation in *Paramecium* produced an "incipient sexuality" – that the "female" progeny of the pairing were supposed to reproduce asexually with vigor, while the "male" reproduced little or not at all.[29] Jennings and Lashley found that to the contrary, conjugation actually made the progeny of a pair resemble each other more in vitality and reproductive power than they resembled others. In other words, there was true inheritance in *Paramecium:* both parents affected the behavioral characteristics of their offspring. In their next study, Jennings and Lashley showed that biparental inheritance also affected size. "As a result of conjugation, the progeny of the two members of the pairs become more alike in size – just as our previous study had shown them to become more alike in their rate of reproduction."[30] As the progeny of one pair became more alike, they began to differ from the progeny of other pairs. Thus diversity was produced in a population, providing a basis on which natural selection could operate.

Lashley's dissertation and related research also contributed to Jennings's effort. The dissertation was supposed to settle the chief problems of heredity and variation in *Hydra.* Lashley examined *Hydra* clones, families descended from single individuals, and found that variations within the clone were not hereditary but were produced by environmental fluctuations. In his discussion of this research, Darryl Bruce notes that this result was confirmed by a subsequent study.[31] But he neglects to mention that this was not the only question Lashley investigated. In fact, Bruce ignores the first part of the dissertation, in which Lashley asked whether hereditarily diverse "races" of *Hydra* existed and concluded that indeed they did. Given Lashley's establishment of a segregated classroom in his graduate school days, this may be the most striking line of questioning in the dissertation. Lashley asked:

> Given a character which is comparable in different individuals, do races, hereditarily diverse with regard to this character, exist? If there are such races, in what characters do they differ and are they numerous or few? Is

[29] Jennings and Lashley, "Biparental Inheritance and the Question of Sexuality in *Paramecium*," *Journal of Experimental Zoology* 14 (1913): 393–466.
[30] Jennings and Lashley, "Biparental Inheritance of Size in *Paramecium*," *Journal of Experimental Zoology* 15 (1913): 198.
[31] Darryl Bruce, "Lashley's Shift," 30.

there inheritance of individual differences within the population, and if so, what part do diverse races play in this inheritance? . . . What is the origin of the diversities between races?[32]

Although within a family descended from a single individual there were no hereditary variations, "the evidence seems to prove conclusively that some internal, hereditary factor caused the differences between [families]."[33] While he admitted that the environment did play a role in creating some of this variety, Lashley remained firmly committed to the idea that populations of *Hydra* did consist of races with different hereditary constitutions:

> It has been found that within a wild population of *Hydra viridis* there are hereditarily diverse races which differ in their number of tentacles at separation from the parent, in their size at a different age, and less certainly in other characters. The differences between such races are permanent so long as the races are kept under the same environment. The evidence favors the view that the differences are truly genotypic. . . .[34]

These results, Lashley wrote, were in accord with the work of Jennings and others, which "leaves no doubt that the existence of diverse races within the species is a general condition in all phyla."[35]

In his 1920 book, *Life and Death, Heredity and Evolution in Unicellular Organisms,* Jennings summarized a decade's work in inheritance in the Protista. Here he drew out the implications of Lashley's dissertation work for the human species: hereditary racial differences did exist, both in the Protista and in human beings. His book, he said, was not about the biology of individuals but about the biology of races. His main theme was to determine the purpose of mating in protozoa. Why, in animals that could reproduce asexually, did sexual reproduction also take place, with an exchange of nuclear genetic material? Jennings considered the possibility that mating somehow rejuvenated the lines, so that their vital processes were stimulated. But this was not always the case, Jennings found; after mating some of the lines suffered a loss of vitality. He therefore discarded the theory of rejuvenescence, and concluded instead that sexual reproduction served to increase diversity, so that after mating there was a wider variation in vitality than before. The characteristics produced by sexual reproduction were passed from parent to progeny, so that eventually a number of hereditarily diverse stocks or families, different "races," appeared. Mating in protozoa had two results: biparental inheritance and hereditary diversity, the one serving to make the offspring of a pair more alike, the other serving to make them different from the

---

[32] Lashley, "Inheritance in the Asexual Reproduction of *Hydra,*" *Journal of Experimental Zoology* 19 (1915): 165.

[33] Ibid., 181.

[34] Ibid., 190.

[35] Ibid., 203.

offspring of different pairs. Thus Jennings claimed that he could actually see evolution occurring. Reproduction brought about diversity in the population, so there could be no such thing as a uniform species.

From the protozoa to man was one small step for Jennings. As diverse races of protozoa existed, so did different races of human beings. As protozoa tended to mate like with like, so as to widen these racial divides even further, so did man also practice "assortative mating."

> Assortative mating is common, too, in higher animals and man. It is well known how strong a reluctance there is in man for strikingly different races to mate; a reluctance that is reenforced by all sorts of social and legal regulations (which regulations, of course, are manifestations of the biological characteristics of organisms). In the case of blacks and whites among human beings, for example, an observer from Mars, examining in the United States the two stocks objectively, would find that in the overwhelming majority of cases white is mated with white, black with black – although some exceptions occur.[36]

Jennings's point was to show that both structural and behavioral characteristics were hereditary. He examined not only such traits as length of spines and size of shell, but also the ways in which those organisms lived. Some strains multiplied rapidly, others slowly; some were hardy and survived easily under adverse conditions, others were delicate; some were active, others quiet; some were adapted to one set of conditions, others to a different set. The continuity between Jennings's earlier work on the behavior of the lower organisms, on their "biography and perhaps psychology," was clearly continuous with the research he carried out with Lashley's assistance. Here he was simply investigating the genetic basis of the behaviors he had so closely observed before.

Bruce writes that "the direct intellectual influence of Jennings on Lashley is difficult to determine," though it seems that Lashley "absorbed Jennings' opposition to vitalism and to the ideas of Jacques Loeb."[37] Clearly, however, Jennings's perspective informed Lashley's much more specifically and more thoroughly than that. Lashley took Jennings's attitude toward consciousness: Lashley never dismissed consciousness as unscientific, but instead believed that it had to be reduced to its physicochemical basis (see Chapter 2). Jennings stressed the spontaneity of the lower organisms, arguing that "the organism is activity" and that the motivation for its behavior came from within, not from its environment. Lashley, likewise, came to believe that the organism's nervous system was in almost constant activity, that stimuli never fell on a static system, that the brain was always organizing its environment (Chapter 4). The continuity that Jennings saw

---

[36] Jennings, *Life and Death, Heredity and Evolution in Unicellular Organisms* (Boston: The Gorham Press, 1920), 192.
[37] Bruce, "Lashley's Shift," 30.

between the regulative behavior of protozoa and human intelligence appeared also in Lashley's analogies between basic biological functions and the highest psychological processes. Lashley freely compared the regulative behavior of the developing embryo to the brain's compensatory capacity (Chapter 6). Like Jennings, he concluded that the source of behavior and intelligence was internal and ultimately genetic. Lashley adopted Jennings's strategy of holding the environmental factors constant in order to investigate genetic diversity (Chapter 9).

These beliefs remained with him from his graduate school days to the end of his career; Lashley always remained much closer to Jennings's outlook than he ever came to Watson's.

2

Lashley, Watson, and the Meaning of Behaviorism

Lashley and Watson

While he was working under Jennings's direction, Lashley began a fruitful collaboration with John B. Watson. Twelve years Lashley's senior, Watson had been trained in comparative psychology at the University of Chicago by Henry Herbert Donaldson and James Rowland Angell. He had initially been attracted by the iconoclasm of the physiologist Jacques Loeb, but Donaldson and Angell dissuaded him from doing his Ph.D. work with Loeb. Instead, Watson wrote his dissertation on the correlation between brain growth and learning ability in rats. For several years afterwards he taught psychology at Chicago.[1]

In 1908, Watson became professor of psychology at Johns Hopkins, and the following year, when his immediate superior James Mark Baldwin resigned, Watson was promoted to the senior professorship in psychology.[2] By then he was already beginning to formulate a materialist position in psychology, which reached full expression in his 1913 behaviorist manifesto, "Psychology as the Behaviorist Views It," published in the *Psychological Review*. "Psychology, as the behaviorist views it," Watson wrote, "is a purely objective, experimental branch of natural science. . . ."[3] Unlike his teachers Donaldson and Angell, Watson believed that psychology could become a real science only by focusing on the study of behavior and ceasing its attempts to determine the content of the human mind.

Regardless of how revolutionary his new psychology actually was, Watson did envision behaviorism as a radical break from all previous psychological theory. Whether following the functionalist school of Donaldson and Angell, or the structuralist school of Edward Bradford Titchener, psychologists had relied

---

[1] Philip J. Pauly, *Controlling Life,* 172–177.

[2] Pauly, "G. Stanley Hall and his Successors: A History of the First Half-Century of Psychology at Johns Hopkins," *One Hundred Years of Psychological Research in America: G. Stanley Hall and the Johns Hopkins Tradition,* eds. Stewart H. Hulse and Bert F. Green (Baltimore: The Johns Hopkins University Press, 1986), 20–51. On Watson's life, see Kerry W. Buckley, *Mechanical Man: John Broadus Watson and the Beginnings of Behaviorism* (New York: Guilford Press, 1989).

[3] John B. Watson, "Psychology as the Behaviorist Views It," *Psychological Review* 20 (1913): 158. For a critical discussion of behaviorism in its cultural context, see David Bakan, "Behaviorism and American Urbanization," *Journal of the History of the Behavioral Sciences* 2 (1966): 5–28.

largely on the introspective method to discover the content of the human mind. By this approach, a normal adult human being was trained to report on his own mental activity. For Watson, introspection could never form the basis of a science because its results could never be confirmed or falsified; only the person introspecting could judge the accuracy of his own observations. The knowledge produced by introspection, therefore, would be completely subjective. A real science, Watson argued, would produce results that were verifiable by an outside observer.

Moreover, when it came to abnormal and especially comparative psychology – Watson's field – introspective psychology could have nothing to say. The "animal mind," if such a thing existed, could not be trained to introspect. As a result psychologists were left with futile analogies between animal and human consciousness and vague guesswork about what the animals were thinking. Jennings too had recognized this problem, and had remained agnostic on the question of animal consciousness; but Watson dismissed the concept out of hand. He concentrated instead solely on the behavior of organisms rather than on the workings of their putative minds. "One can assume either the presence or the absence of consciousness anywhere in the phylogenetic scale without affecting the problems of behavior one jot or tittle; and without influencing in any way the mode of experimental attack upon them."[4]

Recent reinterpretations of Watson's "behaviorist revolution" have shown that his focus on behavior was actually part of a broad movement in psychology in the first decades of this century to create a definitive split between psychology and philosophy.[5] By removing mind from psychology, behaviorist psychologists hoped to carve out a realm of study for themselves in which they would be free from the encroachments of philosophers. Indeed, by 1909, Watson had begun a campaign at Hopkins to separate psychology from the department of philosophy and to make it an independent department.[6] At the same time, he wished to strengthen its ties to biology: he maintained an alliance with Adolf Meyer, the psychiatrist and psychobiologist, and his courses were cross-listed with those of the zoology department. It should not be at all surprising, therefore, that Jennings encouraged his zoology students to take Watson's courses; that Lashley, while still Jennings's student, undertook a very productive collaboration with Watson; or that Jennings arranged postdoctoral support for Lashley to continue his work with the behaviorist.

[4] Watson, *Behavior: An Introduction to Comparative Psychology* (New York: Henry Holt and Company, 1914), 4–5.
[5] John M. O'Donnell, *The Origins of Behaviorism: American Psychology, 1870–1920* (New York University Press, 1985). See also Franz Samelson, "The Struggle for Scientific Authority: The Reception of Watson's Behaviorism, 1913–1920," *Journal of the History of the Behavioral Sciences* 17 (1981): 399–425. On the predecessors to Watson, see John C. Burnham, "On the Origins of Behaviorism," *Journal of the History of the Behavioral Sciences* 4 (1968): 143–51.
[6] Pauly, "G. Stanley Hall," 36.

Lashley and Watson worked together from 1911 to about 1916, and then again in 1918 – before, during, and after Watson's initial formulation and refinement of his behaviorist manifesto. In these early stages of behaviorism, Watson combined and integrated elements which he later perceived as incompatible. Not only did he emphasize the close association between psychology and biology, he also treated heredity and environment, instinct and habit, as interdependent aspects of be- havior, and he stressed the need for both laboratory and field studies in the development of behaviorism. In his post-1920 versions of behaviorism, having left psychology for advertising, Watson separated psychology from biology, disavowed the notion that heredity had any effect on behavior, and gave up fieldwork for laboratory studies of infants. But during the 1910s, Watson's be- haviorism was biological and hereditarian; both his work with Lashley and Lashley's own work under Watson's supervision during those years demonstrate the early compatibility of these ultimately opposing perspectives.[7]

Despite his long and productive collaboration with Watson, Lashley was never a behaviorist in the Watsonian sense. In fact, in the early 1920s, he wrote his own behaviorist manifesto, which differed in crucial respects from Watson's concur- rent formulations of behaviorism. But because of Watson's emphasis in his pre-1920 behaviorism on biology and heredity, Lashley found his own developing views in harmony with Watson's. Together Watson and Lashley worked on field studies of animal behavior and on laboratory studies of the human salivary reflex. Lashley also made significant contributions to Watson's 1914 book *Behavior,* the first extended treatment of his behaviorist position.

The first study that Watson and Lashley published together was their "Notes on the Development of a Young Monkey."[8] This article was essentially Lashley's diary of his observations of the appearance of instincts in the infant monkey. "There is . . . no evidence to show," they concluded, "that the infant monkey ever gained a new activity by imitation. Walking, climbing, eating and even the different vocal sounds appeared as instinctive acts which were merely perfected by practice."[9] Watson's pre-1920 behaviorism, as Hamilton Cravens has noted, depended on the idea of innate reactions – instincts – which were matured by practice.[10] The emphasis on the monkey's instinctual apparatus was compatible with Jennings's belief in the internal control of the behavior of the lower organ- isms. By attributing the young monkey's behavior to the unfolding of instincts, Lashley and Watson were simply doing for a higher mammal what Jennings was doing for the protozoa.

[7] Hamilton Cravens has shown that Watson's pre-1920 behaviorism was hereditarian in "Behaviorism Revisited: Developmental Science, the Maturation Theory, and the Biological Basis of the Human Mind, 1920s-1950s," *The Expansion of American Biology,* eds. Keith R. Benson, Jane Maienschein, and Ronald Rainger (New Brunswick: Rutgers University Press, 1991), 133–163.

[8] Watson and Lashley, "Notes on the Development of a Young Monkey," *Journal of Animal Behavior* 3 (1913): 114–139.

[9] Ibid., 139.

[10] Cravens, "Behaviorism Revisited," 133–163.

The papers that Lashley published on his own following this were also focused on the problem of instinct. The same year, 1913, he studied inarticulate sounds in the parrot, attributing their reproduction to the parrot's courting instinct.[11] The following year, he contributed to the same journal a brief note on the persistence of the sucking instinct in adult cats. This study, Lashley concluded, showed the applicability of animal behavior studies to man, "particularly with respect to the experimental study of the roles of heredity and environment in the development of character."[12] In his work with both Jennings and Watson, then, Lashley emphasized the hereditary component of behavior and tried to distinguish it from the environmental.

But Watson removed his attention from center to periphery, from the brain to the reactions of the sense and motor organs. In this he differed most substantially from Jennings and Lashley. While Jennings spoke of "internal sources of motivity," Watson believed that the sources of stimulation were ultimately environmental. Responding to these environmental stimuli, the organism would make externally detectable movements. Attributing important effects to the central nervous system was just as unscientific as relying on introspection to discover the content of consciousness. The brain was unapproachable by Watson's behavioristic methods; its reactions were unmeasurable. Like consciousness, it was too mysterious to play a role in the new science. As a result, Watson wrote in 1914, "there are no centrally initiated processes."[13] In this he was directly opposed to Jennings, for whom all reactions were centrally initiated.

Watson's shift in emphasis from the central to the peripheral also made enemies of the neurologists. In 1915, for example, the neurologist C. Judson Herrick wrote to Meyer, Watson's colleague, that the rise of behaviorism had struck a blow to neurology. "I judge that you too are somewhat discouraged regarding the outlook in neurology in America. It is not roseate. . . . [W]hen Watson insists upon applying his own rather narrow measuring stick in all other fields to the exclusion of other work and methods, he only makes himself ridiculous."[14] Meyer responded that Watson indeed ignored all consideration of functions in which central nervous activity was important, protesting in his own inarticulate way: "[I do not] eliminate the brain. You evidently do not either, while Watson does. Why should the brain not remain a brain, even if we consider it of importance to realize its resting on solid and live ground?"[15]

[11] Lashley, "Reproduction of Inarticulate Sounds in the Parrot," *Journal of Animal Behavior* 3 (1913): 361–66.

[12] Lashley, "A Note on the Persistence of an Instinct," *Journal of Animal Behavior* 4 (1914): 294.

[13] Watson, *Behavior,* 18.

[14] Herrick to Meyer, March 20, 1915, Adolf Meyer Archive, Series III, Unit 212, Folder 13, Alan Mason Chesney Medical Archives, The Johns Hopkins Medical Institutions, Baltimore.

[15] Meyer to Herrick, December 7, 1915, Adolf Meyer Archive, III/212/13, JHU. On Meyer's views on psychology and psychiatry, see *Defining American Psychology: Correspondence Between Adolf Meyer and Edward Bradford Titchener,* eds. Ruth Leys and Rand B. Evans (Baltimore: Johns Hopkins University Press, 1990).

Though he appeared to neurologists to be rendering their science unnecessary, Watson actually envisioned close alliances between behaviorism and the other biological sciences. In his 1914 book, *Behavior: An Introduction to Comparative Psychology,* Watson explained how the behaviorist was supposed to work closely with biologists. "There should follow a constant interchange, among these sciences, of behavior material, data on evolution, neural structure, and physiological chemistry."[16] Training in behavior had to be accompanied by training in histology, physiology, and experimental zoology.[17] Likewise, "[s]imple working conceptions of what goes on in the nervous system when habits are formed will be very stimulating to the student of behavior."[18] Watson believed behaviorists and neurophysiologists should work together, the one establishing habits in animals, the other correlating them with structural changes in the nervous system.

Watson's purpose in his 1914 book was to reduce behavior to its most basic component, the reflex arc. Reflexes, neural connections between sensory receptors and muscles, underlay all behavior, whether instinctive or habitual. In instinct, the pattern and order of these reflexes was hereditary; in habit, they were acquired. But both instinct and habit depended on the presence of innate modes of response. In habit, these reflexes were at first unorganized: the stimulus called forth numerous reflex responses, from which the proper ones were then selected and fixed through repetition. Innate response, therefore, was important for all behavior, and there was a close relationship between what was hereditarily endowed and what was shaped by environment. The responses in habit were innate, but they depended on the environment for their maturation and perfection.

Watson believed that his reflex conception could be applied to the behavior both of the lowest organisms and of human beings. Even the highest psychological processes involved overt behavior; abstract thought, for example, was always accompanied by slight movements of the tongue and larynx which could be detected, measured, and in turn reduced to reflex components. Thus even the most seemingly complex process was subject to a behavioristic treatment; and thus all behavior, even the most complex, was dependent on innate responses. "It seems safe to conclude that all of the vocations are probably at bottom dependent upon particular hereditary types of organization, i.e. dependent upon the presence of random activity of proper kinds."[19] From this reservoir of hereditary response, the most adaptive sequence of reflexes was culled, just as in the simple habits.

Nevertheless, Watson was careful to avoid oversimplifying his conception of behavior and thereby falling prey to Jennings's criticism of Loeb. Reflexes were not necessarily simple: "the term reflex should include not only the more definite and fixed types of reflexes with which we are already familiar through our studies

---

[16] Watson, *Behavior,* 55.
[17] Ibid., 55n.
[18] Ibid., 51.
[19] Ibid., 186.

in physiology, but also those which are less constant, and less predictable. . . ."[20]
He noted that Jennings, Mast, and Yerkes had shown that reflexes, even in the
protozoa, were not "absolutely fixed and wholly stereotyped forms of reaction."[21]
Unlike Loeb, Watson had no intention to reduce all responses to reflexes of a
stereotyped kind. While he admitted that the same organism might respond in the
same way to the same stimulus, Watson also showed his allegiance to Jennings's
position by noting that "[t]he organism is constantly changing."[22] The external
environment of the organism might be controllable, but not the "internal physiological processes which are also essential parts of the total stimulation and
which also affect markedly the state of the effector."[23] Watson therefore managed
to avoid falling into Loeb's trap of emphasizing "invariability and predictability,"
and at the same time showed his closeness to Jennings's conception of behavior.
Rather than carry Loeb's emphasis on prediction and control unaltered into the
realm of mammalian behavior, Watson wanted to effect a compromise between
his position and Jennings's.

Watson believed that his investigations into the reflex as the basis of all behavior had to take place in both the field and the laboratory. He countered the
ethologists' criticism that behaviorists ignored the activities of animals in their
natural habitats, arguing that field and laboratory studies must complement each
other. This complementarity is especially clear in studies that Watson pursued
with Lashley.

Their most extensive field study of animal behavior was their fieldwork on the
nesting and homing behavior of the noddy and sooty terns. These birds bred in
great numbers on the tiny, barren island of Bird Key, in the Dry Tortugas off the
gulf coast of Florida. Earlier Watson had been to the island and found that the terns
were able to home with great accuracy over hundreds of miles of open sea. To test
this, Watson had Lashley drive the birds to Galveston, Texas, release them, and
observe whether they could find their way home. Lashley subsequently spent six
weeks on Bird Key trying to determine the sensory factors involved in the terns'
ability to recognize their nests. Lashley argued that the study of proximate nest
recognition would illuminate the birds' ability to home over great distances. He
described how he changed the appearance and location of the nests, and how the
birds reacted; his report was essentially a detailed ethological study. Like Watson,
Lashley concluded that the birds showed no evidence of possessing mind or self-
consciousness: "In subjective terms, they show little or no evidence of 'ideational
processes' in their activities." Their behavior had to be understood without recourse to anthropomorphism.

[20] Ibid., 107.
[21] Ibid.
[22] Ibid.
[23] Ibid., 107–8.

At the same time, Lashley underscored the similarity between the birds' behavior and human behavior. When their nests were altered, the birds displayed the same reactions that a human being would under comparable circumstances. But the birds reacted fully, in a "peculiarly impulsive" way, to whatever stimulus was immediate; while the human reaction would have been limited to verbal behavior.[24] In other words, what the bird actually did, the human being would only consider doing. But for the behaviorist, thought processes in human beings were simply discernible and measurable movements of the larynx. Thought was therefore a kind of action; and so the bird and the human being were essentially responding in the same way. Lashley reinforced Watson's contention that the psychology of human beings and of unintelligent animals could be treated identically, and that neither involved mind.

Watson and Lashley also worked together in the laboratory. In 1916, for example, under Watson's direction, Lashley published two studies of the human salivary reflex and its conditioning. Working with human subjects, Lashley designed an apparatus to measure saliva flow in response to various stimuli.[25] Lashley noted that although the data were physiological, some of the reactions he observed were so complex as to "place the topic of reflex secretion very near the borderline between the sciences of physiology and psychology."[26]

By 1916, however, the observation and measurement of externally detectable behaviors had begun to lose its appeal for Lashley. He met Franz and became increasingly interested in the neurological basis of behavior. Watson, meanwhile, had moved his laboratory to the Phipps Clinic of the Johns Hopkins Medical School. He gave up fieldwork entirely and began work on conditioning the reflexes of human infants. Lashley and he still remained on good terms, though Rosalie Rayner, Watson's graduate assistant, became his collaborator on the infant studies. In 1919 Watson published *Psychology from the Standpoint of a Behaviorist,* an application of his behavioristic views to the human case, in which he still drew on Lashley's work. For example, Lashley had designed an apparatus to measure the tongue movements of people involved in some numerical calculation or trying to remember the words to a song. He had found that their tongues and vocal apparatus did indeed make discernible movements. Watson used this experiment to prove that thinking was "subvocal speech," and could be objectively observed, thereby obviating the need for introspection.[27]

[24] Lashley, "Notes on the Nesting Activities of Noddy and Sooty Terns," *Neuropsychology of Lashley,* 23.

[25] Lashley, "Reflex Secretion of the Human Parotid Gland," *Neuropsychology of Lashley,* 28–51. Also Lashley, "The Human Salivary Reflex and Its Use in Psychology," *Psychological Review* 23 (1916): 446–464.

[26] Lashley, "Reflex Secretion of the Human Parotid Gland," *Neuropsychology of Lashley,* 28.

[27] Watson, "Is Thinking Merely the Action of the Language Mechanisms?" *British Journal of Psychology* 11 (1921): 87–104.

But by 1919 there were important changes in Watson's behaviorist position. In addition to giving up field studies, Watson also became increasingly concerned to defend the autonomy of psychology. While in 1914 he had emphasized the cooperation necessary between psychologists and biologists, by 1919 he announced that behaviorists could do their work without regard to the progress of the biological sciences. They were not dependent on physiology but had their own subject matter and their own practical goals to pursue. While the other sciences focused on isolated "parts," the behaviorist was concerned with predicting and controlling the behavior of the whole organism. Psychology, he wrote, is "the study of the organism . . . put back together again and tested in relation to its environment as a whole."[28] Earlier Watson had followed Jennings in noting that the internal physiological processes of the organism had an important effect on its outward behavior. Now he drew a distinction between those internal states and processes and outward reactions, and redefined behavior to focus entirely on the latter. Jennings defined behavior as any of the organism's activities; Watson argued that behavior only properly referred to its outward reactions.

At around this time, Watson became embroiled in an extramarital love affair with Rayner; the resulting scandal forced him to resign from his professorship at Hopkins. He and Rayner relocated to New York, where Watson joined the J. Walter Thompson advertising agency, moving rapidly through its ranks to become vice president by 1924. He remained connected to academic psychology, however, even without his own laboratory. He and Rayner continued to write about child behavior and became popular authorities on the subject. He gave evening lectures on behaviorism at the New School for Social Research in New York, and remained on the editorial boards of several psychological journals.[29]

But by 1924, when Watson published his next detailed exposition of behaviorism, his position had undergone a significant change, probably under the influence of his new career.[30] While before 1920 he had stressed the hereditary component of behavior and the cooperation between heredity and environment, now he argued that behavior was actually entirely environmentally determined. Like the advertising man, who tried to predict and control the actions of people through the manipulation of stimuli, the behaviorist should also focus his attention on the external conditions surrounding the organism. By changing the environment, the behaviorist could shape an organism's behavior to his specifications. Watson

---

[28] Watson, *Psychology from the Standpoint of a Behaviorist* (Philadelphia: J. B. Lippincott, 1919), 20.

[29] Watson, "John Broadus Watson," *A History of Psychology in Autobiography,* Vol. 3, ed. Carl Murchison (Worcester, MA: Clark University Press, 1936), 271–81. See also Ruth Leys, "Meyer, Watson and the Dangers of Behaviorism," *Journal of the History of the Behavioral Sciences* 20 (1984): 128–151.

[30] On the connections between Watson's new career and his radical environmentalist stance, see O'Donnell; and Kerry Buckley, *Mechanical Man: John Broadus Watson and the Beginnings of Behaviorism* (New York: Guilford Press, 1989).

expressed this radical environmentalist position in a famous passage from his 1924 book:

> I should like to go one step further now and say, "Give me a dozen healthy infants and my own specified world to bring them up in, and I'll guarantee to take any one at random and train him to become any kind of specialist I might select – doctor, lawyer, artist, merchant-chief, and yes, even beggar-man and thief, regardless of his talents, penchants, tendencies, abilities, vocations and race of his ancestors."[31]

In the next sentence he acknowledged the extremity of his position: "I am going beyond my facts and I admit it, but so have the advocates of the contrary and they have been doing it for many thousands of years."[32] But his shift in attitude was indisputable. Where once he had seen heredity as the basis for all learning, innate reflexes as the basis of habit, he now saw environment as the maker and shaper of all behavior. Inborn response disappeared from his behaviorism.

Lashley and Behaviorism

This transition to radical environmentalism marked the final parting of the ways between Watson and the biological tradition represented by Jennings and Lashley. In order to reassert his own beliefs and establish his independence from Watson, Lashley wrote his own behaviorist manifesto, which appeared in the *Psychological Review* in 1923.[33] The problem with Watson's behaviorism, Lashley wrote, was not that it had gone too far, but that it had not gone far enough: that it had foundered on the details of its experimental method and had not taken its formulations to their logical conclusion.

Lashley believed that Watson, in defining psychology as the study of behavior, had not found an adequate way to deal with the mind. Behaviorists ruled the mind out of their scientific purview, and as a result were forced to take one of three possible routes. They could argue that although mind in fact exists, it can only be approached by introspection; or they could declare mind incapable of any description, by introspection or otherwise; or they could claim that mind did not exist – that it was behavior, with nothing left over. The first two options, Lashley felt, were nothing but psychophysical parallelism: they supported the untenable idea that mind and body ran in two parallel but nonintersecting tracks. In both cases behaviorism would become nothing but a supplementary system to introspective psychology; it would not be able to account for the data gathered by introspection. The third option, strict behaviorism, had never been fully developed; Watson had

---

[31] Watson, *Behaviorism* (New York: Norton, 1924), 104.
[32] Ibid.
[33] Lashley, "The Behavioristic Interpretation of Consciousness," *Psychological Review* 30 (1923): 237–272, 329–353.

been too concerned with following objective methods and had never managed to confront the introspectionist on his own grounds.

Instead, Lashley argued, behaviorists must stop trying to deny that introspective data were worthy of scientific consideration, and figure out some way to account for them on mechanistic or physiological principles. Parallelism was unacceptable; if behaviorism was to reform psychology, it must reform the whole of psychology, from the simplest behavior to the most abstract thought process. All must be equally explicable in the language of biology, chemistry, and physics. While Watson's main purpose had been to find objective methods for psychology, Lashley believed that questions of method must be subordinate to the development of a mechanistic account of consciousness. "To me the essence of behaviorism is the belief that the study of man will reveal nothing except what is adequately describable in the concepts of mechanics and chemistry, and this far outweighs the question of method by which the study is conducted."[34] While Watson argued that introspection was inappropriate in a scientific psychology, Lashley not only believed that introspective data must be accounted for, he even relied on introspection for some of the evidence in his article. While Watson believed subjectivity had to be eliminated from psychology, for Lashley the real enemy was mind-body dualism; and it seemed to him that in Watson's formulation, behaviorism was coming dangerously close to dualism.

In the interest of maintaining the independence of psychology from biology, Watson paid only cursory notice to what was actually going on in the nervous system. He hypothesized that reflex connections were established, but he never made any attempt to trace such anatomical connections. The fact that the organism responded in a certain way was proof enough of their existence. In 1919, he offered a "general caution" about the role of physiology in the study of behavior:

> While we wish to emphasize the importance of the central nervous system, we do not wish to make a fetish out of it. . . . The main fact about the central nervous system is that it affords a system of connection between sense organs and glands and muscles. Interrupt the pathway at any place and the organism no longer acts as a whole; some phase of the behavior pattern will drop out.[35]

Lashley, on the other hand, believed that the ultimate goal of the sciences was their unification, and so focused his behaviorism on reducing psychology to biological, chemical, and physical principles. When the behaviorist denies that consciousness exists, Lashley wrote, he denies not the existence of certain phenomena, such as memories, thoughts and feelings, but only the conclusion that those phenomena constitute a unique category of things inexplicable on physical

[34] Lashley, "The Behavioristic Interpretation of Consciousness," 244.
[35] Watson, *Psychology from the Standpoint of a Behaviorist,* 153–4.

principles. For Lashley, the purpose of behaviorism was to correlate consciousness with the workings of the nervous system. While Watson shied away from too
great a dependence on the biological sciences, Lashley wrote that "[w]ithout
physiology behaviorism can make but little progress, for its explanatory principles
are physiological and no sharp line can be drawn between the two sciences."[36] He
attributed the weakness of behaviorism to the "backwardness of the science of
physiology" and declared that "[t]he behaviorist's chief handicap is the lack of an
adequate physiology upon which to base his science."[37]

For Watson the value of behaviorism lay in its practical relevance. Its goal was
"the prediction and control of behavior"; it was supposed to have direct implications for society. He intended behavioristic principles to be put to work immediately: by businessmen in training better workers, by educators in producing
better students, by mothers in raising better children. Watson's second career as an
ad man and popularizer was not so much a break with his academic scientific
career as a fulfillment of his earlier promise that behaviorism would yield practical results. In fact, the practical relevance of Watson's behaviorism insured that its
methods would remain objective; if the behaviorist was supposed to be telling
people how better to live their lives, he had to get out on the street and start
observing their behavior. Introspection could not lead to practical improvements;
a subjective science was for Watson fundamentally a useless science. But society
was Watson's laboratory: by focusing on how people actually behaved, his
methods would be objective and his science relevant to real life.

By contrast, Lashley's version of behaviorism was not explicitly practical.
Though he said that one of behaviorism's strengths was that it made of psychology
"a true science of human conduct" – that it did not leave "the problems of everyday life to the 'applied sciences' of sociology, education and psychiatry" – he
himself was unwilling to adopt the goal of practical relevance.[38] While Watson
advocated the prediction and control of behavior, Lashley believed that behaviorism should "discover the wells of human action," finding the origin of a person's
interests, abilities, or prejudices but not taking an active role in modifying them,
as Watson argued. In Watson's 1919 book practicality was the constant refrain: a
chapter called "The Organism at Work" argued that better ventilation of the
workplace would make workers more efficient, and that the use of tobacco and
alcohol would decrease efficiency.[39]

Lashley, on the other hand, never used examples from everyday life in his
article and never applied behaviorism to the solution of social problems. He might
try, for example, to translate a feeling of hunger into its neurological and physiological correlates; but usually he made specific provision to keep his reduc-

[36] Lashley, "The Behavioristic Interpretation of Consciousness," 349.
[37] Ibid., 351.
[38] Ibid., 348.
[39] Watson, *Psychology from the Standpoint of a Behaviorist* (1919).

tionistic behaviorism away from practical problems. When he needed to illustrate his behavioristic principles, rather than applying them to an actual situation, he might imagine a machine, equipped with sense organs, motor organs, and a system of connections between them.[40] This imaginary machine would be able to perform all of the actions we would expect from a conscious human being, but of course would not possess a soul or mind or be conscious in any sense. It would, therefore, be indistinguishable from a human being, and so there was no point in assuming that human beings must possess an immaterial consciousness. While Watson made the same argument by focusing on the actual behavior of human beings and explaining it without recourse to mind, Lashley preferred to perform a thought-experiment.

Lashley kept his science apart from the real world in another way as well. A common argument against the reduction of psychology to physiology is that it fails to capture the quality of experience; that a neurological explanation of the sensation of the color red always falls short of truly expressing what it is like to see redness. The "raw feel" of the experience is missing from the description of chemicals and electrical signals, and there would never be any way to bridge this gap. "There is a persistent demand," Lashley wrote, "that the scientific description shall be capable of arousing the experience of the thing described."[41] He countered this demand by drawing a strict separation between art and science: the raw feel of red, or a memory of childhood, belonged in the realm of art, along with all other qualitative experiences, human aspirations, and social values. Behaviorist psychology should not be expected to reproduce those sensations merely because it sought to explain them in terms of the brain's biology and chemistry. They belonged to one category of experience, scientific explanation to another.

By creating these two incommensurable categories, Lashley evaded the mental qualia argument against his position. He also separated science entirely from social value; science could give us neither the meaning of life, nor the values according to which society should be organized. "It is only by divorcing itself from metaphysics and values and adopting the phenomenological method of science that psychology can escape the teleological and mystical obscurantism in which it is now involved."[42] Scientific explanation became for Lashley entirely descriptive, never prescriptive.

While Lashley's distinction between science and value helped to keep psychology away from practical concerns, it might be criticized for introducing a duality into his militantly antidualistic scheme. After all, the behaviorist was supposed to take all the evidence of mind, including the introspectionists' reports, and reduce everything to neurology and chemistry. Nothing was supposed to fall outside his

[40] Lashley, "The Behavioristic Interpretation of Consciousness," 330–6.
[41] Ibid., 347.
[42] Ibid.

explanatory capacity; and yet here Lashley was in effect admitting that there were certain things that would never be amenable to scientific description.

Lashley and the Mind-Body Problem

While Watson the behaviorist dismissed the mind-body problem as a nonproblem, Lashley the behaviorist spent his career trying to solve that problem. Even after he dissociated himself from behaviorism in the mid-1920s, Lashley still kept as his goal the reduction of mind to biology. His reductionism was tempered by what might be called a holistic perspective; but neither his reductionism nor his holism were of the conventional sort. Reductionism is usually defined as the breaking up of large, complex processes into smaller, simpler elements. Watson's behaviorism, for instance, was classically reductionistic in that it built up complex behavior out of simple reflex connections. Holism, meanwhile, is the idea that certain processes resist resolution into simple components; that properties like life or mind might be considered inviolable wholes which emerge at higher biological or psychological levels.

Lashley rejected both reductionism and holism in these conventional formulations. For him, mind was pattern: "Consciousness consists of particular patterns and sequences of the reactions interacting among themselves and the attributes of consciousness are definable in terms of the relations and successions of the reactions."[43] Mind might be composed of a series of reflex connections, but it made no sense to speak of mind as residing in any one of those connections, or in any single compartment of the brain. "We may speak of an element of consciousness but not of a conscious element."[44] Mind emerged from this system, as one reflex response led to another, and as the succession and integration of elements produced behavior. However holistic this formulation may sound, Lashley rejected conventional holism as a variant of vitalism – the idea that an ineffable spirit or soul gave life and mind to the organism. The holists, Lashley felt, unnecessarily added to the mechanistic account "the conception of *organism* in physiology, and of *personality* in psychology: wholes which are more than the sum of their parts."[45] They spoke of "the organism as a whole," substituting "empty names" for rigorous physiological description.[46]

Lashley argued that consciousness meant nothing more than " 'such and such physiological processes are going on within me.' "[47] He was reductionistic in believing that consciousness was physiology and nothing more. But in emphasizing the importance of pattern, he avoided the pitfalls of a simple-minded reduc-

[43] Ibid., 341.
[44] Ibid., 248n.
[45] Ibid., 270.
[46] Ibid.
[47] Ibid., 272.

tionism. In fact, Lashley's solution to the mind-body problem was rather vague; he seems to have believed that it consisted in finding analogies between patterns of behavior or experience and physiological patterns:

> The essence of consciousness is a field of many elements, organized after the plan of human experience. . . . We must now examine this organization in greater detail to discover in how far it conforms with the types of organization discovered by the physical and biological sciences within their realms of investigation.[48]

Lashley spent the rest of his career looking for correlations between psychological and physiological patterns, and trying to find a middle ground between what he felt were the equally oversimplified extremes of reductionism and holism. Both of the controversies in which he later became involved demonstrated his opposition to the conventional formulations of these philosophies. He argued against the psychobiology of the Chicago school because it relied on irreducible concepts like life, mind, and consciousness. At the same time he turned against behaviorism, arguing that it focused on the elements of behavior, the reflex connections, rather than investigating the whole behavior in its context.[49]

Lashley was trained in a place in which both of these approaches to the organism, the holistic and the reductionistic, were being developed, at a time before they had substantially diverged. Thus during the 1910s Lashley could believe with Jennings that behavior was ultimately under internal control, but still work closely with Watson in the development of behaviorism. I have argued that Lashley's perspective on psychology was thoroughly informed by the work of Jennings, and that Lashley's collaboration with Watson succeeded because, in this early period, Watson's outlook was still close to Jennings's.

---

[48] Ibid., 261–2.

[49] Lashley's attempt to chart a middle course for behaviorism, between the extremes of reflex atomism and a holistic focus on global behavior, was just one among many in the late 1910s and 1920s. In 1922, for example, Edward C. Tolman announced his "New Formula for Behaviorism" in the pages of the *Psychological Review* (Vol. 29., No. 1, pp. 44–53). Tolman thought that Watson's behaviorism was in danger of degenerating into " 'muscle contractions and gland secretions' " – into "mere physiology" – and losing sight of the total behavior pattern (p. 45). Like Lashley, Tolman advocated saving the concept of consciousness, the evidence of mind provided by introspection. Like Lashley, Tolman was interested in the problems posed to behaviorism by imagery, feeling-tone, language, and introspection; he did not dismiss them, but redefined them in terms of their influence on behavior. "Immediate conscious feels" thus became "behavior cues"; the "meaning" of such a cue became the "behavior object." Tolman sought to include in his behaviorism both "the facts of gross behavior and those of consciousness and imagery" (47). "[T]his new behaviorism," Tolman wrote, "will be found capable of covering not merely the results of mental tests, objective measurements of memory, and animal psychology as such, but also all that was valid in the results of the older introspective psychology" (46–7). Though their approaches were different, Tolman's stated intention to improve on Watson's behaviorism and his emphasis on behavior patterns (rather than the individual elements which made then up) were quite similar to Lashley's. See also Laurence D. Smith, *Behaviorism and Logical Positivism* (Stanford, 1986), 80–84.

In one important respect, however, Lashley was closer to Watson than he ever was to Jennings. By the late 1920s, Lashley's two mentors began to differ more thoroughly over the issue of the predictability of behavior. Watson defined behaviorism as the prediction and control of behavior; these aims became even more possible once he had made environment the sole determinant of behavior. The neo-behaviorists, particularly Clark Hull, adopted prediction and control of behavior as their aims as well. Jennings too was certainly interested in control, as his participation in the eugenics movement shows. But he separated himself from behaviorism by adopting a version of holism that denied absolute predictability.

According to Jennings, holism meant that new qualities emerged at each higher level of organization, with the creation of new wholes, and that these new qualities could not have been predicted merely by examining the individual components of the whole.[50] In this way the whole was really more than the sum of its parts: the properties of life and mind which emerged at higher levels of biological organization could not be predicted from studying cells or chemicals in isolation.

Jennings explained this concept by asking the audience at his lectures to imagine the "fiery mist" that existed at the beginning of the world; could the appearance of life at later periods ever have been predicted from it?

> Confronted with samples of that material [the fiery mist], perceiving the particles of which it was composed, their velocities, directions of motion, and distribution in space, could the investigator hope, by logic and the aid of an unlimited computing machine, to compute from its properties and distribution the later emergence of life, including varied sensations, feelings, ideas and thoughts as they have later come forth?

Jennings himself resoundingly answered the question:

> No! From computations that begin with particles, velocities, directions, distributions, there can emerge by computational and logical treatment only particles, velocities, directions, distributions – never a sensation. A sensation of red does not logically or computationally follow from an arrangement of particles in combination with vibrations of a certain wave length. Not the hardiest mechanist has attempted so to derive it.[51]

This newness at each level, this unpredictability, was a central feature of holism for Jennings; it meant, in fact, that biology, psychology, and sociology were granted true autonomy from chemistry and physics, that organisms and groups of organisms had to be studied each at their own level in order to determine their properties. The naturalist was therefore vindicated in his studies; his observations

---

[50] See, for example, Jennings, *Some Implications of Emergent Evolution* (Hanover, New Hampshire: The Sociological Press, 1927). See also *The Universe and Life* (New Haven: Yale University Press, 1933).

[51] Jennings, *The Universe and Life*, 29.

of the behavior of the *Paramecium* or of the sooty tern would never be replaced by the analysis of a physiological chemist. The notion of emergent properties was, Jennings proclaimed, "the Declaration of Independence for biological science."[52] If new properties constantly emerged at successively higher levels of organization, Watson's ideal of complete predictability was also thrown into serious question.

Lashley reinforced Watson's concern with predictability through the emphasis on pattern in his 1923 article. His solution to the mind-body problem was to look for correlations between psychological and physiological patterns; to see how patterns of behavior were related to patterns of nervous function. But pattern implies repetition, and repetition is of course an essential feature of predictability. When Watson sought to establish conditioned reflexes in infants, he too was looking to discover and establish repeated patterns of behavior. The search for pattern was, then, an important aspect of the search for control. "For, after all, when I say that I am conscious of something," Lashley wrote, "I say merely that there exist certain organizations of entities – sensations, images, ideas – describable patterns, the elements of which are indescribable."[53] And again: "It is not alone the attributes of the elements of content but the particular variety and pattern of them that makes up the supposed uniqueness of human consciousness."[54] Lashley's interest in embryology and in Gestalt psychology were also to arise from this concern with pattern; it was a concern that allowed him to avoid the dangers of reductionism, but that also linked him even more securely to Watson's program. "Scientific explanation might be called the manipulative interpretation of the universe," Lashley wrote in 1923. Manipulation, prediction, and control – usually thought to be foreign to Lashley's sensibility – were actually part of his concern with pattern.

On the issue of predictability, Lashley has often been portrayed as much closer to Jennings than to Watson. Darryl Bruce wrote that while Watson wanted to control behavior, Lashley was interested in understanding it; Lashley himself said that he wished "to discover the wells of human action."[55] This distinction between understanding and control, knowledge and power, has helped to maintain Lashley's image as a pure scientist, uninterested in the consequences of his discoveries or in applying his results. How Lashley himself constructed and maintained this image will be the subject of the next chapter.

[52] Jennings, *Some Implications of Emergent Evolution,* 5.
[53] Lashley, "The Behavioristic Interpretation of Consciousness," 340–1.
[54] Ibid., 332.
[55] Bruce, 34; Lashley, "The Behavioristic Interpretation of Consciousness," 348.

3

The Pursuit of a Neutral Science

Prologue

Since Lashley's death in 1958, historians of psychology have done little to modify his own assessment of the history of psychology and his role in it. They imply that in so highly technical a field as neuropsychology, the aim of which is to describe the neural basis of consciousness, there is room only for the most disinterested of truth-seekers. That is certainly what Lashley considered himself, and historians have not challenged his portrayal. Lashley has become a symbol to psychologists of a perfectly neutral scientist.[1] He was, according to himself and others, led to his conclusions purely by induction from his experimental results; free of preconceptions, he never hesitated to tear down any theory, including his own. Indeed, after his brief early endorsement of behaviorism, he never became an ardent follower of any psychological movement, preferring to point out their weaknesses and their conflicts with what he considered the facts. From the mid-1920s onward, Lashley rejected a collection of theories, and by the 1950s he was rejecting the notion that any theory could ever explain the complexities of psychology.

Lashley's opposition to theory has become legendary among psychologists.[2] He himself stressed it in his papers and addresses – writing, for example, that "[a]lways the question, How? punctures the bubble of theory, and the answer is to be sought in analysis and ever more analysis."[3] He perceived his role as oppositional, his scientific style as "negative": he argued *against,* never *for.* In a typical article he would present a multitude of experimental results, but conclude only

[1] See, for example, Sharon E. Kingsland, "A Humanistic Science: Charles Judson Herrick and the Struggle for Psychobiology at the University of Chicago," *Perspectives on Science* 1 (1993): 445–477.

[2] See Frank A. Beach, "Karl Spencer Lashley, June 7, 1890–August 7, 1958," 162. Beach describes Lashley as a "[f]amous theorist who specialized in disproving theories, including his own." See also Edwin G. Boring, "Lashley and Cortical Integration," *The Neuropsychology of Lashley,* xi–xvi. Boring writes: "Or, if he [Lashley] were to insist that he were an agent and not a creator, then he was the agent of nature, for the impressive thing about the papers in this volume is the way in which discovery leads speculation, and not speculation discovery" (xvi).

[3] Lashley, introduction to Heinrich Klüver's *Behavior Mechanisms in Monkeys* (Chicago: University of Chicago Press, 1933), x.

that they disproved certain hypotheses. His findings made nothing certain, he claimed: rather, they made a host of ideas uncertain.

For example, in an article published in 1924, he wrote: "The conclusions which may be drawn from these experiments are wholly negative."[4] Likewise, in his famous address "In Search of the Engram," he concluded: "This series of experiments has yielded a good bit of information about what and where the memory trace is not. It has discovered nothing directly of the real nature of the engram."[5] Nor was this an attitude he reserved solely for public appearances; he maintained it in his private correspondence as well. To Clark Hull, he wrote in 1945: "I realize that this is not a very convincing explanation of the results, but I do not feel under obligation to construct an adequate hypothesis for such a complicated situation."[6] He could only point out the conflicts between his evidence and everyone else's theories, and again and again he noted the destructive aspect of his work. He wrote only one book during his career, because he felt that his experiments were too inconclusive to support any large theoretical system.

Lashley's distance from theory was matched, again both in his own portrayal and in accounts by others, by his seeming detachment from the social and political consequences of psychology.[7] He was unwilling to present his opinions on society and on the contributions of science to society; he seems to have believed that the two belonged to entirely different realms. This is in striking contrast to most of his contemporaries. Both John B. Watson and Adolf Meyer certainly considered the scientist a social expert; Lashley's colleagues at the University of Chicago, C. M. Child and C. J. Herrick, did not hesitate to draw comparisons between the growth of organisms and the development of democratic societies. Moreover, the foundations that supported Lashley's research had explicitly social purposes: the Behavior Research Fund in Chicago, for example, was supposed to aid juvenile delinquents. Yet the fund supported Lashley's work on rats, and Lashley himself resisted the notion that his work had any immediate practical applications. In a letter he noted, "We should not deceive ourselves with the belief that the solution of society's problems can come from science. Science can affirm means but not ends. . . ."[8] For Lashley, science was about understanding, not control; about knowledge, not power.

---

[4] Lashley, "Studies of Cerebral Function in Learning V," *Neuropsychology of Lashley,* 128.

[5] Lashley, "In Search of the Engram," *Neuropsychology of Lashley,* 501–502.

[6] Lashley to Hull, October 1, 1945, Kenneth Spence Papers, Box M938, Archives of the History of American Psychology, University of Akron.

[7] See Kingsland, "A Humanistic Science," 464. Darryl Bruce writes that Lashley wanted to understand neurophysiology, while Watson wanted to control behavior. See Bruce, "Lashley's Shift," 34.

[8] Lashley to Roger J. Williams, June 14, 1946, quoted in Darryl Bruce, "Integrations of Lashley," *Portraits of Pioneers in Psychology,* eds. Gregory A. Kimble, Michael Wertheimer and Charlotte White (New Jersey: Erlbaum, 1991), 321. Lashley continued: ". . . and my pessimism makes me believe that our values will always be dictated by those who can play on the emotions, the poets, preachers, spellbinders, the Huey Longs and the Cardinal Newmans."

Lashley made it appear that his oppositional style of science was the inevitable result of a lifetime of painstaking experimental work. But his image as an objective truth-seeker, innocent of both theoretical and practical considerations, was one that he carefully constructed in the course of his long career.

Lashley's Early Training with Franz

The pivotal event in Lashley's intellectual development during the 1920s was his rejection of behaviorism. This event was explained by Lashley himself – an assessment in which psychologists and historians have since concurred – as the triumph of his own inductive scientific method. His empirical work convinced him that behaviorism, in particular its stimulus-response notion of brain function, could not be true. In its place he argued that the brain worked as an equipotential system, that is, that mental function was a product of the whole brain. But an examination of Lashley's rejection of behaviorism suggests a different interpretation of the episode.

After Lashley received his Ph.D. in 1914, he began a collaboration with the psychologist Shepherd Ivory Franz.[9] At the Government Hospital for the Insane in Washington, D.C., Franz trained patients with brain injuries to resume normal lives, a process which he called "re-education." In order to instill in the patient an attitude of hope, Franz stressed the similarities between the brain-injured and the normal. He believed, for example, that brain-injured patients could be re-educated in the same way that students were taught. In fact, throughout his 1923 book, *Nervous and Mental Re-Education,* the analogy between doctors and teachers became virtually an identity; he wrote: ". . . re-education is to the abnormal what education is to the normal – it is a matter of the acquisition of habits that will enable the individual to take his place in the working, playing, social world."[10] In the preface to the same work he observed:

> It would be best if the patient could be called something other than "patient," something that would always convey to him, to his family, and to his friends, that he is being fitted to take up his life work again. "Pupil," "learner," "scholar," and the more recently used "trainee" are useful but not good. Similarly with the term for the re-educator, who may be a physician or surgeon, but who frequently is not a medical practitioner. If in America the term "doctor" did not so commonly connote a medical practitioner, this term would be the most suitable. In place of this we are forced for the moment to use the convenient "teacher," "instructor," and "physician."[11]

[9] On Franz and his laboratory, see John A. Popplestone and Marion White McPherson, "Pioneer Psychology Laboratories in Clinical Settings," *Explorations in the History of Psychology in the United States,* ed. Josef Brozek (Lewisburg: Bucknell University Press, 1983), 216–222.

[10] Franz, *Nervous and Mental Re-Education* (N.Y.: MacMillan, 1923), 17.

[11] Ibid., vi–vii.

Franz's work had parallels to behaviorism. Like the behaviorists, Franz believed that all learning is the formation of habits, the forging or breaking of connections between stimuli and responses. Because a change of environment could affect what new habits were formed, re-educability was not dependent on some native capacity, but was rather a matter of exposure to the right stimuli. Watson, in fact, believed that Franz's results lent support to the behaviorist conception that human nature was malleable; in his 1919 book, Watson wrote: "Our interest here [in Franz's work] . . . centers around the possibility of reëstablishing certain basal habits connected with the care of the person (eating, drinking, talking), and those connected with simple occupations in individuals who have suffered grave cortical lesion. A few years ago the thought of retraining a paralytic of long standing would not have been seriously entertained."[12]

Lashley may have sought out Franz to work with because of the similarities between Franz's outlook and Watson's. From Franz, Lashley not only learned surgical technique but also acquired a belief in the re-educability of patients. In a 1920 article in *Psychobiology,* Lashley wrote: "The recent studies of reëducation in hemiplegia, aphasia and apraxia show that the loss from cerebral lesions is never necessarily permanent in man and that an unlimited though slowly acquired vicarious functioning is possible."[13]

Behaviorist Neurology

In 1920, after settling at the University of Minnesota, Lashley began the series of experiments on the brains of rats that made his reputation as a meticulous researcher. Using the surgical method he learned from Franz, Lashley destroyed portions of the rats' brains and, after allowing the rats to recover, tested them for their ability to learn and remember a series of tasks. The tasks included threading mazes of varying complexity, tripping the latches on problem boxes, and discriminating between different patterns. He found that by and large, despite missing parts of their brains, the rats had not lost these abilities.

From his experiments Lashley concluded that the brain functioned as a whole: that there were no discrete brain compartments specific to certain functions. To this view Lashley adhered for the rest of his life. Functions were not localized in the brain; rather, they were a property of the whole brain.[14] Mental function required a certain critical amount of brain to be present, but it did not matter what specific parts remained intact. Lashley summarized his findings in two principles

---

[12] Watson, *Psychology from the Standpoint of a Behaviorist* (Philadelphia, J.B. Lippincott, 1919), 298.

[13] Lashley, "Studies of Cerebral Function in Learning," *Psychobiology* 2 (1920): 126.

[14] Lashley did believe in gross anatomical localization – that is, that certain areas of the brain were specific to a certain class of functions (vision, olfaction, etc.) – but he contended that specific visual memories or capacities, for example, were not lodged in specific cells of the visual cortex. Vision was a property of the whole visual cortex, just as intelligence was a property of the whole cerebral cortex. See, for example, "In Search of the Engram," *Neuropsychology of Lashley,* 491.

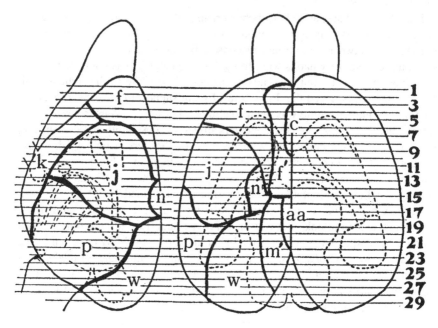

Chief cytoarchitectural areas of the rat's cortex. (Reprinted from *Brain Mechanisms and Intelligence.* Used with permission of the publisher.)

f, f′ and n = motor or pyramidal region
k = olfactory
j = somesthetic
p = auditory
w, m′, and aa = visual

The dotted lines represent subcortical structures. The horizontal lines, each numbered (on the right) from the front to the back of the rat's cerebrum, are a convention Lashley used to identify and locate landmarks in the cerebrum.

that became synonymous with his name: equipotentiality (that all parts of the brain are equally capable of carrying out all functions) and mass action (that only a certain critical amount of brain is necessary for its proper functioning).

Like Franz, Lashley harmonized Watson's behaviorist theory with his own experimental results. As Franz used the behaviorist idea of learning-as-habit to reinforce his belief in re-educability, Lashley also combined behaviorist notions of brain function with his own theory of equipotentiality. According to behaviorism, learning – habit-formation – was the establishment of connections between stimulus and response, the laying down of reflex arcs that would traverse the cortex and link sensory and effector organs. Responses could become associated with certain stimuli by the conditioned reflex method, which Watson and other

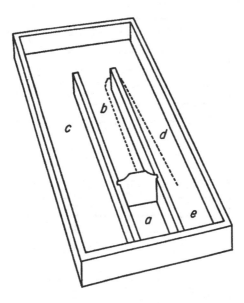

The simple maze that Lashley used in early studies: *a* is the starting compartment, *e* is the food, *d* is the alley leading to the food, *b* is the central alley, and *c* is a cul-de-sac. The dotted line shows the shortest path to the food, the path taken by well-trained animals. (Reprinted with permission from *Neuropsychology of Lashley.*)

behaviorists envisioned as the creation of actual pathways in the brain and nervous system: once the pathway was laid down, the connection was forged, and the response was established as a habit.

Lashley subscribed to this view of brain function, which he adopted from Franz and Watson, and saw it as perfectly compatible with the notion of an equipotential brain. In his 1920 article, Lashley wrote that "the reflex connections involved in habits may be laid down in any part of the cortex." He continued: "the only essential condition for learning is the simultaneous activity of two reaction systems which are in anatomical connection by association fibers." There were no cerebral areas responsible for learning, no localization of functions: "The investigations of all the functional anatomy of the rat's brain . . . have all pointed to an almost complete interchangeability of function among the different parts of the cortex, and a rapid recovery from the effects of cerebral operation."[15]

Lashley's harmonization of equipotentiality with the reflex theory of brain function is crucial to understanding his construction of the image of the objective scientist. In 1926 Lashley began to argue that the reflex theory was actually

[15] Lashley, "Studies of Cerebral Function in Learning" 102, 124, 115.

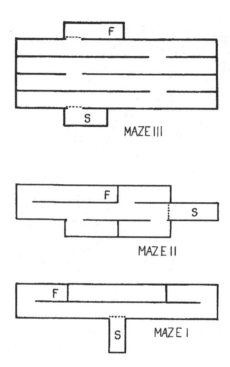

Ground plans of the enclosed mazes that Lashley used in his 1929 work *Brain Mechanisms and Intelligence*. (Reprinted from *Brain Mechanisms and Intelligence* with the permission of the publisher.)

incompatible with the principle of equipotentiality. If there were reflex paths traversing the cortex, he reasoned, then surgery would destroy them, and his experimental rats would be incapacitated. But the rats were not incapacitated; therefore there could be no reflex paths mediating the habits. Before 1926, the fact that the rats were unimpaired convinced him that the surgery had not interrupted the reflex paths. After 1926, the lack of impairment convinced him that there were no reflex paths at all.

Throughout the 1920s, however, Lashley was doing the same experiments and getting the same results, and even interpreting them in the same way: the brain must function as an equipotential whole because destruction of isolated portions did not impair its functions. At first he saw equipotentiality as compatible with, and even explained by, the reflex theory. Later he argued that the reflex theory was contradicted by his evidence of equipotentiality. By 1926 Lashley considered the reflex theory of brain function to be synonymous with the doctrine of localization: "reflex connectionism" became interchangeable with the notion that ideas could

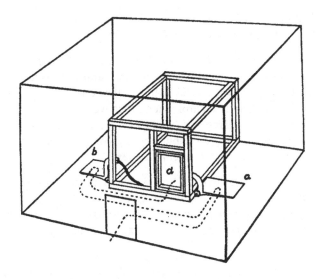

The double platform box that Lashley used in studies throughout the 1920s. The rat must press down the platforms *a* and *b* in order for the door *d* to open. (Reprinted from *Brain Mechanisms and Intelligence* with permission of the publisher.)

be localized in specific brain compartments; and both theories were for him equally bankrupt.

Toward the end of his life, Lashley recounted that he had turned against the reflex theory because it was fundamentally a localizationist doctrine, and his experimental results had convinced him that there could be no localization of function in the brain. But this account represents Lashley's reconceptualization of the role of experiment in the development of his theory. Essentially, Lashley rewrote his story so that he would appear to have rejected a theory on the basis of empirical evidence. But neither his experimental approach nor his results changed markedly during the course of the 1920s. Yet in 1920 he used the reflex theory to support equipotentiality and reject localization, and in 1926 he rejected the reflex theory along with localization in favor of equipotentiality. In 1920 he conceived of brain function in terms of stimulus-response pathways, and in 1926 he completely rejected this conception. If his experimental findings were no different, then how did the change come about?

Rat Brain Ablation in the 1920s

The experiments that Lashley did early in his career demonstrate the compatibility of reflex principles with the theory of the equipotential brain. In 1917, he and

Franz published in *Psychobiology* a study of habit formation and retention in the rat after cerebral destruction. They concluded, in part:

> The ability of the animals to form habits after the loss of those parts of the brain which are normally used in learning, the reëstablishment of motor control after the loss of the stimulable area of the cortex and of the corpus striatum [centers thought to control voluntary movement] and the seeming equipotentiality of the different parts of the frontal pole in the functioning of complex habits go far toward establishing the complete functional inter-changeability of all parts of the cerebral cortex.[16]

But their belief in the complexity of brain function, in the notion of equipotentiality, was completely compatible with a behaviorist neurology, the idea that the brain operated on reflex principles. Rather than dismissing the reflex concept as too simplistic to account for the equipotential brain, Lashley and Franz adapted its explanatory power:

> It seems then that for retention some part of the frontal pole must be preserved but no particular part seems necessary. This is, perhaps, what might be expected if we abandon the purely diagrammatic concept of the reflex cortical arc as a simple chain of neurons and consider that for the performance of even a simple movement a number of such arcs are required.[17]

With many reflex pathways involved in mediating a habit, destruction of a portion of the brain would probably not destroy all the pathways; those remaining intact could continue to mediate the habit.

In 1920, also in *Psychobiology,* Lashley published the first of a series of articles on cerebral function in learning. Here he wrote: "The most important neurological concept bearing upon nervous function in learning is that of the reflex character of all behavior." The problem was no longer one of determining where functions were localized in the brain; rather, "every reaction of the organism is carried out by transmission of impulses over reflex paths differing only in the number of cells and the complexity of organization intervening between receptor and effector."[18] In this series of experiments Lashley found that no single part of the cerebrum was more crucial for learning than any other part; the brain seemed "absolutely equipotential," and as much as half the cerebrum could be destroyed without markedly affecting the rat's ability to learn.[19] Again, he concluded from this that "the reflex

---

[16] Franz and Lashley, "The Effects of Cerebral Destruction Upon Habit Formation and Retention in the Albino Rat," *Psychobiology* 1 (1917): 133.

[17] Franz and Lashley, "Effects of Cerebral Destruction," 132.

[18] Lashley, "Studies of Cerebral Function in Learning," *Psychobiology* 2 (1920): 57.

[19] Ibid., 101. In this study he found that retention of visual habits was destroyed by the destruction of the occipital pole of the cortex; yet the rats retained the ability to see, and could relearn tasks dependent on vision as rapidly as normal rats (111–113).

connections involved in habits may be laid down in any part of the cortex."[20] In this study he destroyed not only the motor areas of the cerebrum but the sensory areas as well: he knocked out both components of the reflex arc. But the capacity of the brain to compensate for lost areas was unaffected; rather than proving to Lashley that conditioned reflex arcs did not exist, these experiments gave "some insight into . . . the course of the conditioned-reflex arcs through which learned reactions are mediated."[21]

Two later installments of the studies of cerebral function in learning appeared in 1921 in *Brain* and in 1922 in the *American Journal of Physiology*. The purpose of these two studies was again to destroy both the efferent (motor) and afferent (sensory) areas of the cortex and then to test the animal's capacity to form and retain habits. The conclusion from the 1921 study was that "the supposedly motor regions of the cerebrum do not lie in the direct path of conditioned-reflex arcs and are of only secondary importance in the performance of voluntary movements."[22] In other words, this study was another expression of equipotentiality in terms of reflex arcs. The 1922 study examined the role of the "visual areas" in the forma- tion and retention of habits. These areas, thought to be restricted to the occipital pole of the rat's cortex, were destroyed, but nevertheless habits dependent on vision were relearned. In order to determine whether any other part of the cortex had assumed the visual function, other parts of the cortex were destroyed in a second operation. Despite the subsequent destruction, the habit was still retained, indicating that no other part had assumed the function.

Lashley's conclusion from this experiment was again anti-localizationist: it demonstrated the "diffuse functioning of the cortex," or the principle of equipo- tentiality, that the cortex as a whole mediated habits. He wrote: "Such an equipo- tentiality of function demands the existence of numerous fibers passing to and from all parts of the area, so that each conditioned reflex is mediated by numerous equivalent arcs, conducting impulses which summate for the performance of a given act."[23] Again, it was equipotentiality which was opposed to localization and

---

[20] Ibid., 102.

[21] Ibid., 123.

[22] Lashley, "Studies of Cerebral Function in Learning III: The Motor Areas," *Brain* 44 (1921): 275. In this study Lashley found that the habit of discriminating different light intensities could be formed and retained after destruction of any part of the cerebrum, except for a certain area in the occipital pole called the visual area (276).

[23] Lashley, "Studies of Cerebral Function in Learning IV: Vicarious Function After Destruction of the Visual Areas," *American Journal of Physiology* 59 (1922): 65–6. This study showed that the habit of visual discrimination (in this experiment, learning to choose a lighted and avoid a dark alley) was normally mediated by the visual area in the occipital lobe. When this area was destroyed, the habit was lost but then could be relearned, and was retained after a second operation destroying other parts of the cortex. Thus no one part of the brain took over the visual function from the occipital lobe, and within the occipital lobe itself, visual function was distributed: destruction of a part of the lobe resulted in a general lessening of efficiency of the habit, not the knocking out of specific segments of it.

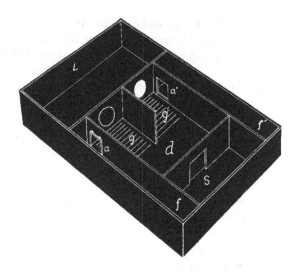

The box Lashley used for training rats in brightness discrimination. The animal starts at *S* and passes through to *d,* where it is faced by the choice of a light or dark alley. *g* and *g'* are electric grills, *a* and *a'* are trapdoors made of celluloid, *f* and *f'* are food compartments. To train the rats, Lashley locked the door in the dark alley and charged the grill. *L* is the light source, which can be moved to illuminate either alley. In use, Lashley kept the entire box covered except for an observation hood above *d*. (Reprinted with permission from *Brain Mechanisms and Intelligence.*)

which was easily harmonized with the notion of reflex arcs. Lashley continued to dispute the localization of functions in the cortex: "Even the fundamental importance of the localized sensory projection areas seems questionable in view of the ease with which sensory function may be recovered after destruction of a normal sensory area."[24] In fact, Lashley used conditioned reflexes to argue specifically against localization. The two concepts were not linked but opposed: ". . . when considered in the light of the reflex concept, much of current interpretation of the cerebral localization of 'mental' functions becomes almost meaningless and little is left save the fact that conditioned-reflex arcs traverse different parts of the cortex."[25] Lashley was clearly not ready to give up the idea of conditioned reflex pathways and sought to explain equipotentiality in terms of them.

In 1924, Lashley was still clearly contrasting localization of function and the reflex theory. He wrote:

The "reflex" conception of cerebral function, although still a theory and notably inadequate to account for all the phenomena of cerebral function

---

[24] Lashley, "Vicarious Function after Destruction," 44.
[25] Ibid.

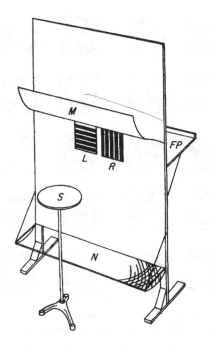

The jumping stand apparatus used for testing discrimination between visual patterns. The rat is placed on the stand *S* and trained to jump through one of the two openings, *L* or *R*, which are covered with squares of cardboard each bearing the patterns to be discriminated. The pattern the rat is being trained not to choose is fixed rigidly; the other will swing open when the rat jumps against it. *M* is a metal shield which will help the rat through one of the openings if it aims too high; *N* is a net which will catch the rat if it falls; *FP* is a food platform. (Reprinted with permission from *Neuropsychology of Lashley.*)

because of oversimplification in its formulations, is too well supported by evidence on nerve conduction and analogy with spinal functions to be disregarded in favor of any speculations concerning the localization of "psychic" functions.[26]

In the same article, Lashley concluded that the destruction of the motor areas brought about no loss in retention of the habit. This did not persuade him that there are no reflex paths that traverse the cortex, but rather that they do not descend from these motor areas.[27]

Perhaps a lack of the right sort of evidence kept Lashley from arguing against the reflex theory. Years later, in relating how he lost faith in the reflex theory,

---

[26] Lashley, "Studies of Cerebral Function in Learning V: The Retention of Motor Habits After Destruction of the So-Called Motor Areas in Primates," *Neuropsychology of Lashley,* 110–1.

[27] Lashley, "Retention of Motor Habits," *Neuropsychology of Lashley,* 132.

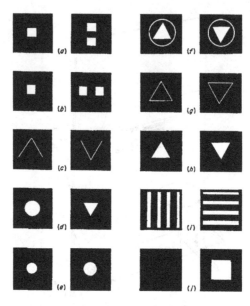

The different patterns between which rats were taught to discriminate. The cards bearing the patterns were six inches square. (Reprinted with permission from *Neuropsychology of Lashley.*)

Lashley said he found that accurate movement was possible without sensory input. This convinced him that reflex pathways did not exist: if sense input was eliminated and the habit was still carried out – albeit less efficiently – there could be no reflex pathways traveling from sense organ to effector. If a necessary part of the reflex was destroyed and the habit was still performed, it was clear that the habit was not mediated by a reflex arc.

But very early in his career Lashley had done experiments involving destruction of both motor and sensory portions of the nervous system, and they had simply confirmed his belief in reflex arcs. In an early experiment, for example, Lashley observed a young man who exhibited, as a result of a gunshot wound to the spinal cord, partial anesthesia of both legs and motor paralysis below the knees. Lashley tested this subject for his ability to control movements of his knee joints in the absence of sensory input from the muscles. He concluded that "accurate movement of a single joint is possible in the absence of all excitation of the moving organs."[28] This conclusion, however, did not lead him to argue against the reflex concept.

[28] Lashley, "Accuracy of Movement in the Absence of Excitation from the Moving Organ," *Neuropsychology of Lashley,* 71. All exteroceptive input was also eliminated – the subject was blindfolded, aural clues were absent, and so forth.

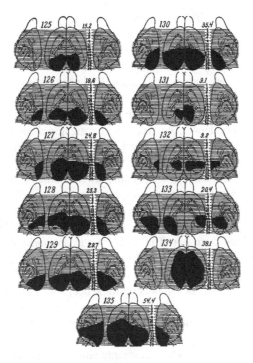

Diagram showing lesions of the occipital (visual) area in the rat. The number to the left is the identification number of the case, that to the right is the percentage of the cortex destroyed. (For explanation of numbered horizontal lines, see caption on page 52.) (Reprinted with permission from *Brain Mechanisms and Intelligence*).

This diagram shows that Lashley extirpated all parts of the visual area; the degree of loss to the visual habit was always proportional to the amount of tissue injured and not to its location.

Lashley used lesion diagrams of this type to argue that no specific problems in maze-running resulted from specific lesions; there was always, rather, a general deterioration in abilities corresponding to the amount of tissue destroyed. This was the case whether or not the animals had the use of their senses. Lashley found that "retention of the maze is unaffected by enucleation of the eyes or by the section of kinaesthetic paths of the cord" (*Brain Mechanisms,* p. 99). Thus, he argued, a rat could run a maze without sensory clues; and inability to run the maze was a result not of simple sensory disturbance, but of a more general deterioration corresponding solely to the amount of brain tissue destroyed.

Redefining "Reflex"

In 1926, Lashley's views underwent an abrupt change. From this point to the end of his career, he became vehemently opposed to the behaviorist idea that reflexes underlay all behavior, even the most seemingly advanced. He also began to associate the reflex theory with the doctrine of localization of function. "Reflex connectionism" and "localization" became virtually synonymous terms in his writings. This striking change in Lashley's science had its source neither in any new findings from his experiments nor even in any new interpretations of his findings, contrary to what has long been believed about Lashley. The change, rather, came from a reorientation in his theoretical position: specifically, from his redefinition of "reflex" behavior.

During the 1920s and 1930s, reflex theory was the subject of a heated and ultimately inconclusive debate. Whether an investigator chose to accept the theory, and in which version, came to depend largely on the settling of a nonempirical, but crucial question: what should "reflex" mean? An investigator had to decide how broad an explanatory concept the reflex should be; to what behaviors it should apply; and what the relationship should be between the "reflex" and the "higher" functions. These questions were not empirical because, simply put, the actions of an animal or a human being did not come with labels identifying which were reflex and which were not. The problem was so murky that by the 1930s three distinct conceptions of the reflex had emerged, each with its own empirical and theoretical supports. Lashley's conversion came about not from empirical work, but rather when he made the nonempirical decision to change his definition of "reflex."

Establishing criteria to demarcate "reflex" from "nonreflex" was no easy task. As a start, one might identify a reflex as an involuntary action, and higher functions as voluntary; but even such a seemingly simple definition led straight to a quagmire. As Franklin Fearing noted in his classic 1930 history of reflex action:

> The broad distinction between voluntary and involuntary is somewhat uncertain in view of the difficulty of defining the term "voluntary." "Automatic" actions are not easy to classify; for the physiologist the term is frequently used with reference only to those reactions that are wholly unconscious, such as the beating of the heart, and glandular activity. On the other hand, the highly learned actions such as playing the piano, typewriting, etc., are called "automatic" by the psychologist. The issues are still further confused by the tendency of modern physiologists and psychologists, especially the latter, to apply the term "reflex" to all types of nervous action which involve a receptor mechanism, an afferent conductor, central elaboration, an efferent conductor and an effector mechanism. The Cartesian idea and the concept of earlier physiologists of the reflex as a specific kind of response distinct from all other types of action is then completely eliminated; unless, of course, all nervous action is presumed to be involuntary, machine-like and inevitable.[29]

---

[29] Franklin Fearing, *Reflex Action: A Study in the History of Physiological Psychology* (Baltimore: The Williams and Wilkins Company, 1930), 286–7.

These difficulties resulted in three distinct conceptions of what "reflex" meant, all current by 1930. According to the strictest definition, which Fearing called the "specific" theory, the reflex was a mechanical and automatic action, not conscious or purposeful, and qualitatively distinguishable from willed or voluntary action. However, certain learned behaviors, such as walking, singing, dancing, bicycle riding, and so forth, could, with enough practice, become involuntary, and therefore reflex; which meant that conscious control was no longer necessary to achieve their performance.

By the second definition, the notion of reflex behavior was broadened somewhat. According to the "genetic" theory, there was a continuous series of behaviors from reflex to willed action. The reflex was still considered a specific type of action, but there was "no sharply drawn line of demarcation between reflex and other varieties of neuro-muscular action."[30] Reflexes were "machine-like" in nature and occurred without consciousness, yet they could be brought under conscious control; likewise, "volitional actions may become involuntary."[31] Fearing placed the neurophysiologist Charles Scott Sherrington in this "genetic" group, and quoted Sherrington's explanation of how a long-practiced skill may lose "conscious accompaniments":

"The transition from reflex action to volitional is not abrupt and sharp. Familiar instances of individual acquisitions of motor coordination are furnished by the cases in which short, simple movements, whether reflex or not, are by practice under volition combined into new sequences and become in time habitual in the sense that though able to be directed they no longer require concentration of attention upon them for their execution. As I write, my mind is not preoccupied with how my fingers form the letters; my attention is fixed simply on the thought the words express. But there was a time when the formation of the letters, as each one was written, would have occupied my whole attention."[32]

According to the third definition, the "mechanical," the reflex was interpreted as the prototype of all nervous action. In this third group Fearing placed Pavlov, Watson, and the behaviorists; they advocated the broadest possible conception of the reflex. No clear distinction was made between reflex and willed action; in fact, attributing certain behaviors to the influence of the "will" was not an explanation the behaviorists considered worthy of a scientific psychology. The characteristics of the simple reflex, as it appeared in lower animals or in a spinal animal, were the characteristics of all behavior, even the most complex and intelligent. All behavior was composed of " 'a sense organ, a sensory neuron, a connecting neuron, a

---

[30] Ibid., 289.
[31] Ibid., 292.
[32] Sherrington, quoted in ibid., 292.

motor neuron, and a motor or secretory organ.' "[33] Watson, Fearing noted, believed that the same simple reflexes, alone or in combination, comprised both instincts and habits, no matter how complex those habits became.

Fearing's discussion of these three points of view demonstrates that by 1930 the reflex was anything but a clearly defined and thoroughly understood entity. Psychologists had borrowed it from physiology in order to lend their study of the mind some scientific respectability, as a tool to sort through a staggering array of psychological phenomena. Instead, the nature of the concept itself became a bone of contention.

What was the "true" reflex? If it was a behavior that was invariable, predictable, mechanical, inherited and stereotyped, then there was the difficulty that such an action hardly ever occurred in the intact animal. If one modified the definition so that a reflex action could be subject to conscious control, then one faced the problem of demarcating between automatic and voluntary actions, opening the possibility that a learned behavior could, with sufficient practice, become reflex. Where was the crucial experiment that could decide these questions, once and for all? In 1930, there wasn't one: such questions were settled partly, but only partly, through observation and experiment. To a very great extent, an investigator simply had to choose where he would draw the line around the reflex concept, how broad a class of behaviors he would label "reflex."

What brought about Lashley's mid-1920s conversion was his decision to narrow the reflex concept. In 1921, Lashley belonged to Fearing's "mechanical" group, along with the behaviorist Watson, believing that habits were mediated by conditioned reflex arcs that traversed the cortex. Long continued practice established the reflex character of these habits even more securely; practice laid down a kind of well-worn path, a path of least resistance, so that when an initial stimulus was presented, the habit would unfold automatically. Lashley did experiments to prove that such conditioned reflex paths did indeed exist in the cortex and did not get reduced to subcortical levels, even after long practice, as some contemporary beliefs held.

To demonstrate just how automatic these habits could become, Lashley gave the example of a pianist who practices a piece until she can carry on a conversation while she plays. The reflex arcs involved in playing the piece become so well established that the action becomes truly involuntary, truly separate from voluntary control and dissociated from a voluntary behavior like speaking. "It is this capacity to function without exciting reaction systems other than those directly concerned with its performance that characterizes the automatic habit," Lashley wrote in 1921.[34] An originally voluntary action has become reflex: the example is

---

[33] A.P. Weiss, quoted in ibid., 299.

[34] Lashley, "Studies of Cerebral Function in Learning II: The Effects of Long-Continued Practice upon Cerebral Localization" (1921), *Neuropsychology of Lashley,* 105.

directly analogous to Sherrington's explanation of the automatic character of handwriting.

Lashley's conversion consisted in his narrowing of the reflex concept so that it could no longer apply to such complex behaviors as piano playing. After 1926, Lashley began to believe that the reflex was a specific type of action, quite distinct from complex and presumably voluntary behaviors; conditioned reflex arcs no longer underlay all behavior. To support this view, Lashley used – interestingly enough – the example of a pianist playing a difficult and rapid passage: the same example he had used in 1921 to argue the opposite position!

In his 1948 address, "The Problem of Serial Order in Behavior," one of his strongest refutations of behaviorism, Lashley argued that the reflex theory could not account for piano playing. His explanation was simple: the pianist's fingers had to move too rapidly for stimuli to direct them. If piano playing were to consist of chains of reflexes, and if reflexes were in turn composed of connections between stimulus and response, then the performance of one element in the series should become the stimulus that provides the excitation for the performance of the next element. The sensation that one note has been successfully played should, on this theory, provide the stimulus that results in the playing of the next note.

But, Lashley argued, this conception is inadequate to account for the playing of a rapid passage. There is simply not enough time between notes for a stimulus to travel from the pianist's eyes to her finger muscles, in order to set off the next reaction in the chain. As Lashley put it in 1948, "The finger strokes of a musician may reach 16 per second in passages which call for a definite and changing order of successive finger movements. The succession of finger-movements is too quick even for visual reaction time."[35] Such rapid motor patterns must be independent of sensory controls; they must be centrally supervised. Such was Lashley's argument in 1948.

Note that what has changed here is Lashley's conception of the reflex, not his empirical example, the observation of piano playing. In 1921, Lashley imagined the playing of music to be an automatic habit, mediated by conditioned reflex arcs traversing the cortex. After 1926, Lashley believed that the playing of music was too rapidly executed for stimuli to play a role in its direction, conceiving the reflex as a chain of stimuli and responses. His early conception of the reflex was much broader than this later one; but both conceptions were perfectly capable of explaining the same facts, and Lashley used the same example to support both conceptions, different though they were.

The main point here is that the example of piano playing, or of any complex behavior, was not the *experimentum crucis* by which Lashley disproved the reflex theory. It did not suddenly occur to Lashley after 1926 to think about piano playing, which realization then convinced him of the inadequacy of the reflex

[35] Lashley, "The Problem of Serial Order in Behavior" (1948), *Neuropsychology of Lashley,* 515.

concept. Lashley had thought about complex behaviors as early as 1921 and, strikingly, had found the reflex concept perfectly able to account for them. It was only when Lashley narrowed the range of the reflex that the complex behaviors fell outside its jurisdiction.

By 1948, Lashley was so thoroughly turned against the reflex theory that his friend the Gestalt psychologist Wolfgang Köhler might have been speaking for Lashley when he said:

> "If I feel a little disappointed in the work of behaviorism, the reason is not so much a certain innocence in its treatment of direct experience and in its imitation of adult physics, but its astounding sterility in the development of productive concepts about functions underlying observable behavior. . . . [I]t is scarcely a satisfactory achievement for the behaviorist to have taken the old concept of reflex action from physiology . . . and to give us no further comprehension into the formation of new individual behavior than is offered by his concepts of positive and negative 'conditioning'. Why should behaviorism be so utterly negativistic in its characteristic statements? 'Thou shalt not acknowledge direct experience in science' is the first commandment and 'Thou shalt not conceive of other functions but reflexes and conditioned reflexes' is the second."[36]

Lashley's Rejection of Behaviorism

It was in the mid 1920s, when Lashley decided to reject the reflex theory and to turn against behaviorism, that Watson began to espouse a radically environmentalist position in behaviorism. While correlation does not necessarily imply causation, the coincidence of these two events is highly suggestive. His elder colleague's newfound environmentalism was, I believe, linked to Lashley's conversion. In seeking to distance himself from what he considered Watson's extreme views about the power of the environment to shape behavior, Lashley jettisoned the entire behaviorist system, including, of course, the concept of the reflex character of all behavior. He also removed all traces of environmentalism from his writings.

Lashley's rejection of behaviorism – the definitive break for which he became so renowned[37] – is a classic case of the underdetermination of theory by evidence.

---

[36] Wolfgang Köhler, quoted in Fearing, 309–310.

[37] At the Hixon Symposium, held at the California Institute of Technology in September 1948, the meeting at which Lashley delivered his paper on "The Problem of Serial Order in Behavior," his colleagues responded enthusiastically to his rejection of behaviorism and the reflex conception of behavior. Commenting on Lashley's paper, the physiologist Ralph Gerard said, "Actually, I find it impossible to think through or even towards the complexities of behavior if restricted to atomic units traveling along atomic fibers." Likewise, the embryologist Paul Weiss remarked, "There are still some authors who try to save the old associationist idea that actually the input shapes the

Neither in 1920 nor in 1926 did Lashley have sufficient evidence to decide for or against the reflex theory; it was a decision that he made on the basis not of evidence but of nonempirical values. In the early 1920s these values pushed him in one direction, toward the reflex theory and behaviorism; in the late 1920s they pushed him in the opposite direction, away from behaviorism. His underlying allegiance was to a hereditarian belief in the determination of intelligence, which was compatible with the behaviorism of the early 1920s, but not with behaviorism as it evolved into environmentalism in the late 1920s. Watson's "radical environmentalist" tract, which appeared in 1924, redefined behaviorism in the mid and late 1920s, thereby losing it followers like Lashley, who believed it had ceased to be a scientifically legitimate theory.

How did Lashley change his mind about behaviorism? In the seventh in his series of studies of cerebral function in learning published in 1926 in the *Journal of Comparative Neurology,* Lashley expressed for the first time an antibehaviorist standpoint, by linking the reflex theory with the doctrine of localization. Both theories, he wrote, argued for a "mosaic arrangement of faculties" in which specific functions were assigned discrete locations in the brain. Just as the localizationists identified individual compartments for each mental power, the reflexologists traced the localized paths through the nervous system from organs of sense to organs of response. In 1926, Lashley rejected both of these hypotheses because they could not explain the dynamic function of the central nervous system which he had observed. But as late as 1924, Lashley had found the reflex theory perfectly capable of explaining the brain's dynamic functioning, and had set it in opposition to the localization doctrine.

In this 1926 article, however, the theoretical landscape had entirely changed: reflexes were dependent, he wrote, on connections between particular neurons, which did not fit the plasticity of function demanded by his data. "Brightness discrimination is not a simple reaction which can be stated readily in terms comparable to the descriptions of spinal or conditioned reflexes."[38] And he concluded in part that "[a]s Herrick says, 'the concept of the reflex is not a general master key competent to unlock all the secrets of brain and mind.'"[39] However, the evidence had remained much the same from his earlier work: he was still finding that learning of new visual habits was not dependent on the part of the cortex thought to mediate visual habits, and that retention of a habit was proportional to the amount of functional tissue remaining intact.

Of course, Lashley himself believed that his evidence drove his theorizing. He

structure of the output. I think they are fighting a losing fight, and I think that today's discussion ought to give them the coup de grace." Both are quoted in Lloyd A. Jeffress, ed., *Cerebral Mechanisms in Behavior: The Hixon Symposium* (New York: John Wiley, 1951), 138, 141.

[38] Lashley, "Studies of Cerebral Function in Learning VII: The Relation Between Cerebral Mass, Learning and Retention," *Journal of Comparative Neurology* 41 (1926): 41.

[39] Ibid., 45.

claimed that after 1926 he turned against the reflex theory because it could no longer explain the evidence of his experiments. In 1929, for example, he wrote that a rat could run a maze perfectly without sensory input or ascending paths in the spinal cord. This offered clear disproof of the reflex theory, in which a chain of sensory signals and motor responses must be established; without the sense input, the chain was interrupted, yet the task was still performed. This, he argued, "implies conceptions of plasticity in nervous function which run counter to the whole doctrine of conditioned reflexes."[40]

According to Lashley's own account of his conversion from the reflex theory, evidence from his experiments had made him change his mind. In 1931 he wrote:

> I began life as an ardent advocate of muscle twitch psychology. I became glib in formulating all problems of psychology in terms of stimulus-response and in explaining all things as conditioned reflexes. . . . And the result is as though I had maliciously planned an attack on the whole system. Accurate movement seemed possible without sensory control . . .[41]

In 1950 he wrote:

> I first became sceptical of the supposed path of the conditioned reflex when I found that rats, trained in a differential reaction to light, showed no reduction in accuracy of performance when almost the entire motor cortex, along with the frontal poles of the brain, was removed.[42]

During the 1920s, then, Lashley's image matured into the one that remained with him for the rest of his life. With his abandonment of behaviorism and its central reliance on the reflex doctrine, he presented himself as a pure scientist, one who lacked allegiance to any theoretical system, one whose ideas about nervous function and behavior were driven solely by the results of his experiments. Looking beyond this image of purity, however, we can see that Lashley turned away from behaviorism not for empirical reasons, not because it had been disproven by his research, but because it no longer harmonized with his hereditarian commitments. As we will see, he maintained these commitments throughout the late 1920s and 1930s, all the while refining his image as a pure scientist.

Lashley and APA Politics

In marked contrast to his claims to purity and neutrality, Lashley was not immune to the politics of his discipline. In fact, he managed to attain positions of power and prestige within the psychological community. In 1929, for example, the year

[40] Lashley, "Spinal Conduction and Kinesthetic Sensitivity in the Maze Habit," *Neuropsychology of Lashley,* 186.
[41] Lashley, "Cerebral Control Versus Reflexology: A Reply to Professor Hunter," *Journal of General Psychology* 5 (1931): 14.
[42] Lashley, "In Search of the Engram," *Neuropsychology of Lashley,* 480.

his relentlessly apolitical monograph *Brain Mechanisms and Intelligence* appeared, he was elected to the Presidency of the American Psychological Association, the central professional society for scientists of mind and behavior. How could an avowedly neutral scientist also be an important player in disciplinary politics? Ironically, it may have been the very forcefulness of his claims to purity that put him in position to ascend the ranks of the APA.

From the mid 1920s through the 1930s, the APA was dominated by an establishment both socially and scientifically conservative. During the 1929 stock market crash and the ensuing Great Depression, the APA's most powerful network – the psychologists Robert M. Yerkes, Lewis Terman, Edwin G. Boring, and Richard M. Elliott – faced an employment crisis among young psychologists of tremendous proportions, as graduate programs turned out Ph.D.s unable to find work as teachers and researchers. According to Lorenz Finison, the psychological establishment adopted a "social darwinist and restrictivist outlook" toward these new members of the profession, a sink-or-swim mentality, which held that the "worthy men" would find work, while those who didn't weren't any good anyway. Boring, for example, actually favored raising standards for entrance into the profession, thereby assuring from the start that only a few new graduates would make it to the job market.[43]

Along with this crisis of employment came a serious split within the APA between basic and applied interests. The establishment favored basic science; Franz Samelson has noted that for ten years, from 1925 to 1935, all of the APA presidents had been "self-designated experimentalists," and all but one were "members of that elite group, the Society of Experimental Psychologists"[44] (founded as a haven for purists like E. B. Titchener, Boring's teacher). Meanwhile, certain liberal factions within the APA argued for trying to expand the employment opportunities of young psychologists by setting up certification programs for those who wished to pursue applied or consulting work. A struggle over certification ensued, from which the establishment emerged the victor. As a result, during the 1930s, several splinter groups broke off from the APA – such as the Psychologists' League and the Society for the Psychological Study of Social Issues – comprising those psychologists with applied, consulting, and sociopolitical interests.[45]

There can be little doubt where Lashley stood on these issues. He was tied not only to Yerkes and Elliott because of the Minnesota connection, but also to Boring, who celebrated Lashley in his 1929 *History of Experimental Psychology*

---

[43] Lorenz Finison, "Unemployment, Politics and the History of Organized Psychology," *American Psychologist* 31 (1976): 749.

[44] Franz Samelson, "The APA Between the World Wars: 1918–1941," in *The American Psychological Association: A Historical Perspective,* eds. Rand B. Evans, Virginia Staudt Sexton, and Thomas C. Cadwallader (Washington, D.C.: APA, 1992), 139.

[45] See Finison, 747–55, and Samelson, 119–147.

and tried during the 1920s to lure Lashley to Harvard. On the unemployment crisis and the basic/applied question, Lashley sided strongly with the establishment. The establishment, in turn, seemed only too happy to promote him within its ranks. From his membership on the APA governing council (the body which for nearly two decades, from the mid 1910s to the mid 1930s, drew up the nomination ballots for president), Lashley was elected APA president in 1929, the direct successor to Edwin G. Boring.

4

Neuropsychology and Hereditarianism

An attempt to relate phylogenetic and individual differences in behavior to brain structure is therefore rather an adventure in correlating the mysterious with the unknown.

Karl Lashley[1]

The Biological Approach to Psychology

Lashley's rejection of behaviorism involved not only a disavowal of the reflex theory, but also an equally strong denial of the importance of environmental influences on behavior. The anti-environmentalist emphasis of his argument is especially clear in his exchange of views with Walter S. Hunter in the *Journal of General Psychology*.

In 1930, Hunter published a criticism of Lashley's concept of cerebral equipotentiality, in which he claimed that Lashley's data did not demand a new theory of neural activity but rather conformed to the reflex arc theory. The central issue was whether behavior was peripherally or centrally controlled; with the behaviorists, Hunter argued that environmental stimuli, acting through the senses, determined behavior, while Lashley claimed that the central nervous system acted independently of the environment. Citing a great deal of evidence to prove that sensory stimulation controlled the various habits that Lashley's rats displayed, Hunter wrote:

> [T]he results show that we are not justified in concluding, as Lashley does, that the maze habit [for example] is not controlled by any stimuli but by a central neural engram which unwinds, the figure of speech is mine, like some Victrola record when the rat is placed in the maze.[2]

In his response published the following year arguing against the reflex concept, Lashley emphasized the isolation from the environment of the nervous processes

---

[1] K. S. Lashley, "Structural Variation in the Nervous System in Relation to Behavior," *Psychological Review* 54 (1947): 326.

[2] Walter S. Hunter, "A Consideration of Lashley's Theory of the Equipotentiality of Cerebral Action," *Journal of General Psychology* 3 (1930): 459.

underlying behavior. The central nervous system, not environmental stimuli, was in control of behavior. He wrote:

> I have objected to the kind of psychology which formulates its theories in terms of S and R, where S represents an environmental situation and R a product of muscular activity, and which assumes that S wholly determines R (or many S's many R's), believing that because its diagrams are simple, neural processes must be equally so.[3]

In denying that the environment entirely shaped behavior, Lashley was denying a main tenet of behaviorism. His experiments during the 1930s and 1940s were really attempts to isolate completely the central nervous system from the influence of the environment. His series of experiments on the rat's visual system involved eliminating all sensory input and testing the rat's ability to carry out tasks. These studies focused on the inherent capacities of the central nervous system alone.

Neuropsychology and the Intelligence Testing Movement

Lashley did his research in the 1930s surrounded by a community of psychologists deeply concerned with the hereditary basis of intelligence and other mental traits. He believed that his studies of rat brains had a direct bearing on the problem, and read and cited the work of the intelligence testers, particularly Edward L. Thorndike, L. L. Thurstone, and Charles Edward Spearman.

In 1931, Thorndike, a psychologist at Columbia Teachers' College, published the series of Messenger Lectures he had given at Cornell University several years before on the subject of human learning.[4] Thorndike's early work is usually regarded as a precursor to behaviorism: in William James's basement, so the story goes, Thorndike taught chicks to solve puzzle-box problems by making the animals associate the right answer with a satisfying effect and the wrong answer with an irritant.[5] Like Pavlov and most of the behaviorists, Thorndike believed that all mental capacity was built up out of connections between stimulus and response; but rather than turn this connectionist concept to environmentalist purposes, as Pavlov did, by the late 1920s Thorndike was building up a case for eugenics. He drew a distinction between his version of connectionism and the Pavlovian or behaviorist version: Pavlov's experiments did not reproduce true learning, Thorndike argued, because in actual learning mere repetition of a stimulus could not create a connection, as Pavlov had thought. Thorndike was therefore a curious

---

[3] Lashley, "Cerebral Control Versus Reflexology: A Reply to Professor Hunter," *Journal of General Psychology* 5 (1931): 16.

[4] Edward L. Thorndike, *Human Learning* (New York: The Century Co., 1931).

[5] Robert Boakes, *From Darwin to Behaviorism: Psychology and the Minds of Animals* (Cambridge University Press, 1984), 68. See also John C. Burnham, "Thorndike's Puzzle Boxes," *Journal of the History of the Behavioral Sciences* (April 1972): 159–167.

mixture, and his case provides proof that belief in heritability of intelligence was compatible with a connectionist theory of neural function.

For Thorndike, connectionism led logically to eugenics. The superior intelligence of human beings was due to the larger number of neural connections formed in the human brain: the qualitative difference between human beings and animals was entirely reducible to a quantitative increase in connections. "And this," he told a Cornell audience in 1929, "may encourage us to believe in the possibility of the future evolution of human intellect and learning," because there was no foreseeable limit to the "number and fineness" of connections that could be formed.[6] At the end of his lectures, Thorndike prophesied:

> Of what sort the learners of the future will be, we do not know, but of the possibility of eugenics in intellect and character there can be no doubt. Human individuals differ by original nature as cats and dogs and tulips and roses do. They differ as truly and probably as much in the genes which determine intellect and character as in the genes which determine height or strength or facial appearance. Intellect is apparently the result not of one or two determiners but of many, so that the task of breeding high ability in or breeding idiocy out is much more complicated and laborious than the task of getting or getting rid of curliness of hair or pinkness of blossom. But it is not impossible. I dare to believe that a time may come when a child born an idiot by a germinal defect will be as rare as a child with twelve toes.[7]

At the beginning of his 1938 book *Primary Mental Abilities,* L. L. Thurstone wrote: "One of the oldest psychological problems is to describe and to account for individual differences in human abilities. How are these abilities and the great variations in human abilities to be comprehended? And just what is an ability?"[8] A pioneer in intelligence measurement, a founder of the journal *Psychometrika* and a contemporary of Lashley's at the University of Chicago, Thurstone devised and used tests to answer these questions, in order to determine not only intellectual capacity but traits of personality as well.

Thurstone concluded that intellect was not reducible to the atomistic stimulus-response connections that Thorndike had advocated, but rather to several "primary abilities" such as a "numerical factor," a "verbal factor," and so forth. In order to isolate these primary abilities Thurstone used the technique of multiple factor analysis, for which he is perhaps best known. He thought of these primary abilities as genetically determined: "There seems to be some evidence for the genetic interpretation of number facility as an inherited trait. Occasionally this

---

[6] Thorndike, 177.

[7] Ibid., 198–9.

[8] L. L. Thurstone, *Primary Mental Abilities* (Chicago: University of Chicago Press, 1938), 1.

ability is found to be extremely conspicuous even at an early age, and it seems then to be more or less independent of other abilities."[9]

In an earlier work, *The Nature of Intelligence* (1924), Thurstone called stimulus-response connectionism "a fallacy" because it concentrated on the environment as the source of motivation rather than on the self, where the focus of psychology properly belonged: "The stimulus is not primarily provocative of living. We, ourselves, are."[10] His deemphasis upon the environment in favor of the capacities of the reacting organism is similar to Lashley's isolation of the central nervous system from the influence of external stimuli; both were antibehaviorist moves. In 1941 Lashley wrote a letter to the Chairman of the Division of Biology and Agriculture of the NRC strongly supporting a grant application from Thurstone.[11]

Most congenial to Lashley's outlook was the work of Charles Edward Spearman (1863–1945), the British statistical psychologist and Grote Professor of Mind and Logic at University College in London. Spearman was the inventor of the two-factor theory of intelligence, in which mental capacity was determined by a general factor ("g") and by a number of specific factors ("s"). Each specific factor was associated with a particular reaction system or mental process, and varied not only among individuals but also among different processes in the same individual. The general factor, however, pervaded all mental processes and held constant in all processes during an individual's lifetime. The factor g, then, was representative of the amount of general mental energy; while s represented the efficiency of the various specific "mental engines."[12] Although g and s were supposed to be separate factors, Spearman never treated them independently: s was really dependent on the amount of g, which pervaded the whole organism.

Spearman's theory, then, was a variation on a one-factor theory, rather than a true two-factor theory (involving two independent variables). General intelligence, g, was the important factor to be measured, itself hereditarily determined and determinative of all an individual's capacities. Spearman found that g was not only the marker of general intelligence, but that it played a crucial role in tasks requiring sensory discrimination, and that it was correlated with memory as well.[13]

For Spearman, one of the most important aspects of this theory was its practical applicability: individuals were to be tested for their quantity of g and then chan-

---

[9] Ibid., 83.

[10] Thurstone, *The Nature of Intelligence* (New York: Harcourt, Brace and Co., 1924), 13.

[11] Lashley to Robert F. Griggs, October 30, 1941, Lashley Papers, Folder "Committee on Human Heredity, NRC," Yerkes Regional Primate Research Center, Emory University.

[12] Charles Edward Spearman, *The Abilities of Man: Their Nature and Measurement* (New York: Macmillan, 1927), 137.

[13] Ibid., 184.

neled into the specific profession for which their intelligence was suited. For example, he wrote that:

> The preceding considerations [about detecting intelligence] are apt to arise on looking at a procession of the unemployed, and hearing the common remark that they are mostly the unemployable. But need they be so necessarily? . . . perhaps every one of these persons could at any rate do something that would make him a treasure in some great industrial concern; and there seems no reason why a few should not have even become famous – in such occupations, for instance, as those of dancer, jockey, or player of a popular game.[14]

Spearman was also clearly interested in finding the physiological or neurological correlate of *g,* a process that Stephen Jay Gould has called the "reification" of intelligence, its transformation into a thing that actually resided in the brain.[15] Spearman envisioned *g* not as being localized in a specific area of the brain, but rather as a property of the whole brain. It is not a coincidence that this is reminiscent of Lashley's equipotentiality theory. Spearman saw in Lashley an ally and used his theory for support; Lashley, in turn, believed that Spearman's work bolstered his own. Both Spearman and Lashley believed that the physiology of the cortex had to be understood if intelligence was to be understood. "Unfortunately," Spearman lamented, "this is just the region where physiology is at present most backward. Even in the most up-to-date textbooks, a very small number of pages are given to it, and these themselves are chiefly filled with anatomy, psychology or mere hypothesis, instead of definite physiological facts."[16]

Like Lashley, Spearman set out the two competing theories of brain activity – localization versus whole-brain functioning – and came down on the side of the latter. Spearman cited a number of examples that echoed Lashley's work: after examining individuals with cerebral damage, he noted that they did not show isolated defects in mental capacity; rather, "the definite experiments and measurements showed that in every case the most obvious cognitive injury was always of the same kind, namely, universal."[17]

Lashley's results easily accommodated Spearman's theory that the whole cortex produces the factor of general intelligence. In his 1929 Presidential address to the American Psychological Association, Lashley said:

> Both the animal experiments and the clinical material point to the conclusion that a given area may function at different levels of complexity, and

---

[14] Ibid., 221.
[15] Stephen Jay Gould, *The Mismeasure of Man* (New York: Norton, 1981), 256–272.
[16] Spearman, *Abilities,* 400–1.
[17] Ibid., 397.

lesions may limit the complex functions without disturbing the simpler ones. Further, we cannot ascribe this limitation to the loss of some necessary elementary functions or to disturbances of nutrition or to shock, for it has been shown in some cases to be solely a function of the quantity of tissue. In this respect the limitation of complexity seems to accord with Spearman's view that intelligence is a function of some differentiated nervous energy.[18]

In his 1931 book *Creative Mind,* Spearman cited this passage triumphantly, noting that his idea that "the mind acts as if it disposed of a fixed amount of energy . . . after long being especially rejected and attacked by physiologists, has now at last found in physiology its most definite support." For Spearman, Lashley's work was "epoch-making."[19]

This, then, was the hereditarian intelligence testing context in which Lashley did his work in the late 1920s and 1930s. His 1929 monograph – the only book he wrote – reflects this context particularly clearly and embodies many of the assumptions of the intelligence testing movement. First, there is the title of the book itself: *Brain Mechanisms and Intelligence.* Like the intelligence testers, Lashley was searching for correlations between the way the brain functioned and the way intelligence was manifested. The discovery of brain mechanisms would provide "positive clues as to the nature of intelligence," and the testing and determination of intelligence, in turn, could suggest possible models of neural function. The psychological and the neurological, then, were interrelated and mutually determinative categories; in 1929 Lashley echoed this idea by dividing his discussion of theory into these two broad classes. As psychologists had concluded that intelligence was either a unified entity or one reducible to many separable constituents, neurologists had also theorized that the brain functioned either as a whole or as a mosaic of independent elements. Lashley believed that his own equipotentiality theory of brain function implied a certain theory of intelligence.

Second, Lashley chose carefully the psychological theories of intelligence that he believed merited "consideration in a deterministic world." Not only did this criterion exclude those theories that were "decidedly animistic in trend" and that fell "wholly outside the scope of scientific treatment" (Herrick's, Tolman's, and McDougall's), but it also omitted any mention of Clark Hull, who had the previous year set out an environmentalist theory of intelligence in his book *Aptitude Testing.* All four theories of intelligence that Lashley did discuss were explicitly hereditarian: the unit factor theory ("developed chiefly by students of heredity");

[18] Lashley, "Neural Mechanisms in Behavior," *Neuropsychology of Lashley,* 202.
[19] Spearman, *Creative Mind* (New York: D. Appleton and Co., 1931), 30. At Lashley's death, a copy of *Creative Mind* was found in his library, its flyleaf bearing the inscription: "*Professor and Mrs. Lashley,* with kindest remembrances and wishes from *The Author.*" This copy is now in the collection of Lashley's books at the Yerkes Regional Primate Research Center. See also Spearman's praise of Lashley in his autobiography, in *A History of Psychology in Autobiography,* vol. 1, ed. Carl Murchison (Worcester: Clark University Press, 1930), 299–334.

Spearman's two-factor theory; Thorndike's aggregate theory; and Koffka's Gestalt hypothesis, which Lashley believed proved that the organization of perception was innate.

But both this introductory discussion and his conclusions about intelligence show that Lashley was firmly in Spearman's camp. At the beginning, though he admitted that "the whole problem is in confusion" and that none of the theories was sufficiently supported by the evidence, he said that "[o]n the whole the psychological evidence seems to favor the existence of some single variable like Spearman's *g*, in addition to special abilities."[20] And in his concluding discussion he invoked Spearman's doctrine by using Spearman's language: "The results of the present experiments lend support to the theory which conceives intelligence as a general capacity." And again: "In this there is close harmony with theories of a general factor determining efficiency in a variety of activities."[21]

Improvement of Intelligence

For Lashley, this debate about how the brain functioned – whether as a whole or as the sum of diverse capacities – turned on the issue of the improvability of intelligence. He made a clear connection between the equipotential brain and the hereditary determination of intelligence: because the brain functioned as a whole, Lashley argued, all of its neurons were involved in all its reactions, and so it had literally no room for improvement. All brains were always operating at the upper limit of their capacity; some would naturally operate better than others, but there was nothing that could be done about it.

Those who were sympathetic to an environmentalist perspective specifically countered this reasoning. The Johns Hopkins psychologist Knight Dunlap, for example, argued that the tremendous excess of unused neurons in the brain permitted improvement or at least modification of the brain's capacities. The Lashley/Dunlap skirmish is a perfect example of two researchers with opposing agendas who looked at the brain and saw two different things.[22] Lashley had evidence that "all the cells of the brain must be in almost constant activity, either firing or actively inhibited."[23] Dunlap, on the other hand, believed that "in every brain there is an excess of neurons over those actually brought into functional operation."[24]

---

[20] Lashley, *Brain Mechanisms and Intelligence: A Quantitative Study of Injuries to the Brain* (Chicago: University of Chicago Press, 1929), 11.

[21] Ibid., 173.

[22] For a similar treatment of an earlier period in the history of the brain sciences, see Steven Shapin, "The Politics of Observation: Cerebral Anatomy and Social Interests in the Edinburgh Phrenology Disputes," in *On the Margins of Science: The Social Construction of Rejected Knowledge,* ed. Roy Wallis (Keele: University of Keele, 1979), 139–178.

[23] Lashley, "In Search of the Engram," *Neuropsychology of Lashley,* 502.

[24] Knight Dunlap, "Psychological Hypotheses Concerning the Functions of the Brain," *The Scientific Monthly* 31 (1930): 106.

For Lashley, the argument against localization and in favor of the equipotential, constantly active brain implied an argument for the status quo, for keeping people in the places that heredity had assigned them. For Dunlap, a vast reservoir of quiescent brain cells represented an opportunity for improvement of capacity and argued against congenital restrictions of ability.[25] In a lengthy and elaborate thought experiment in a 1930 article on the functions of the brain, Dunlap imagined English and African babies swapping brains. Because the brains showed no innate differences, they would take on the characteristics of the bodies into which they were transplanted.

A strange piece of fancy, Dunlap's imagined experiment nevertheless made an important antihereditarian point: physical characteristics might be determined by heredity, but mental capacity was wholly influenced by environment. Each one of us was born with a great storehouse of brain cells, only a small fraction of which we ever use. Theoretically, given the right environment, there would be nothing to prevent us from fulfilling that immense potential, nothing to prevent an African brain from becoming an English one. Of course, in reality, Dunlap admitted, our brains' environments (the peripheral nervous system, the rest of our body, our education and experience) did place limits on how our brains developed. But in Dunlap's image of a modifiable mind, these limits were not innate to the brain.

The issue of improvement of mental capacity is equally clear in another skirmish Lashley had, this one with Pavlov. In 1932, Pavlov published a reply to some of his critics, one of whom was Lashley.[26] Lashley had brought his usual criticisms against reflexology, that it was oversimplified and a hindrance to progress, and had argued that more important to the study of brain function was Spearman's concept of intelligence and an "analogy to the tissue of sponges and hydroids."[27] This analogy was Lashley's way of explaining the phenomena of equipotentiality and mass action: embryonic tissue, like brain tissue, could recover its full function after part of it had been destroyed. This capacity for restitution of function after injury convinced Lashley that the gross anatomy of the brain had very little to do with its ability to function: function seemed not to depend on which structures were present or absent.

For Pavlov, however, this conclusion was absurd: we cannot, he argued, "take for granted some kind of a contradiction between structure and function": the details of structure must sooner or later reveal their functional significance.[28] This linking of structure to function was the goal of Pavlov's conditioned reflex work, while Lashley, he claimed, advocated "[i]ndeed – one can say – some kind of

---

[25] Lashley made this consequence of Dunlap's argument explicit in "In Search of the Engram," *Neuropsychology of Lashley,* 498.

[26] I. P. Pavlov, "The Reply of a Physiologist to Psychologists," *Psychological Review* 39 (1932): 91–127.

[27] Ibid., 101.

[28] Ibid., 104.

bodiless reaction."[29] Moreover, Pavlov explained Lashley's results by arguing that the reflex connections were complexly intertwined, and that Lashley's rats could still perform their tasks because his surgery had not severed those connections: the same argument that Lashley himself had made in the early 1920s.

But the issue at the heart of their disagreement was whether human beings were improvable creatures. Though Pavlov believed that the reflex system that organized human behavior demonstrated the close analogy between human beings and machines, he was clearly interested in keeping a place for modifiability of capacity in his mechanical model. Although reflex connections were the basis for human behavior, these connections could always be altered by a change of environment; and this, in Pavlov's poetic argument, gave reason for hope:

> The chief, strongest and ever-present impression received from the study of the higher nervous activity by our method, is the extreme plasticity of this activity, its immense possibilities: nothing remains stationary, unyielding; and everything could always be attained, all could be changed for the better, were only the appropriate conditions realized.[30]

Equipotentiality and Hereditarianism

We have seen how Lashley's hereditarian conception of intelligence helped to support his idea that the brain operated as an equipotential whole. From the start, the equipotential brain had been a controversial idea, and as Lashley's career progressed, the idea became more, not less, embattled. Lashley, however, remained a staunch opponent of localization, defending the notions of equipotentiality and mass action and never abandoning them. I believe they held too much hereditarian significance for him to have given them up.

Nonetheless, Lashley did struggle with the definition and limits of equipotentiality, and as he confronted new data and mounting evidence against his idea, he worked to explain what it did and did not mean. For example, Lashley made it very clear from the outset that he did not intend equipotentiality to deny the obvious observation that the brain had a structure. Equipotentiality, he emphasized, did not mean that the brain was a formless mass, everywhere structurally (and therefore functionally) the same. The brain was anatomically differentiated: different compartments were clearly responsible for different functions. The striate cortex, for example, was the area of the brain specialized for vision. According to the theory of equipotentiality, though, within these different cortical areas, function was distributed, not localized. Each of the various cortical fields, then, worked as an equipotential whole.

[29] Ibid., 114.
[30] Ibid., 126–7.

One of the best examples of Lashley's ongoing revision and refinement of the concept of equipotentiality was his 1946 article with George Clark called "The Cytoarchitecture of the Cerebral Cortex of *Ateles:* A Critical Examination of Architectonic Studies."[31] Lashley and Clark used two different species of monkey, *Ateles* and *Macaca,* to argue against the use of architectonics – the drawing of cerebral maps – to correlate the structure of the cortex with different mental functions. Against such a strictly localizationist idea, the authors stood firm, conceding a gross anatomical parcellation of the cortex but nothing finer-grained than that: "At present any deductions concerning function of the striate area from the architecture have an extremely tenuous basis," they insisted. A ringing endorsement of equipotentiality followed: "Except for a falling off in acuity and color sensitivity, psychologic data indicate a uniform functional activity in all parts of the striate cortex."[32]

In their article, Lashley and Clark presented two major objections to the theory and practice of architectonics, two objections which if considered in isolation appear to contradict each other. But when these objections are understood in a hereditarian context, they actually complement each other, coexisting in a productive tension rather than in opposition.

The first was the classic antilocalization objection that structure and function could not be so easily correlated. Lashley and Clark admitted that the brain was composed of a myriad different structures, but it was not clear which of these recognizable features of cortical anatomy actually represented a functional differentiation. The structure/function correlation could be pushed only so far; beyond the grossest anatomical level it immediately broke down. As the authors argued:

> The discovery of the striate and Betz-cell areas and of their correspondence to functional divisions of the cortex led to an optimistic search for other cerebral "organs." The early charting of the cerebral surface was done at a time of maximal faith in the separate localization of mental faculties, at a time when Henschen could speculate about the localization of single ideas in single cells, and it required no very convincing evidence to gain acceptance of architectonic parcellations as representing functional divisions.[33]

Again, they claimed, "The problem is not one of charting cerebral 'organs' as Brodmann conceived it, for it is both unlikely that all visible differences in cortical architecture represent differences in function, or that all functionally different fields differ in cellular arrangement."[34] The structure/function correlation was not

---

[31] *Journal of Comparative Neurology* 85 (1946): 223–306.
[32] Ibid., 278.
[33] Ibid., 290.
[34] Ibid., 299.

a reliable one, Lashley and Clark argued, and it proved a major stumbling-block for theories of cortical localization.

Along with this objection to the search for cerebral organs came a corollary. Just as it was mistaken to look for precise structural/functional parallels, it was equally misguided to suggest that these parallels were more striking the higher up the evolutionary ladder one climbed. It was not the case, Lashley and Clark believed, that the rat's brain was a simple undifferentiated mass, everywhere equivalent in function, while a human brain was increasingly functionally differentiated. Rather, the generic rat brain and the generic human brain were very much alike. In their own words:

> That the evolution of the mammalian cortex has entailed the differentiation of more and more areas of diverse function has become a firmly entrenched dogma; man's superior mental endowment is the result of his possession of a greater number of capacities, more rigorously localized in the cortex. . . . From the standpoint of modern psychology, such a view is absurd and, on the basis of comparative anatomy, unconfirmed. The plan of cerebral organization is the same throughout the mammalian series.[35]

For Lashley to admit otherwise would have been seriously to jeopardize his experimental findings and particularly to invalidate their significance for human intelligence. Lashley did nearly all his research on rat brains, and an objection often brought against him was that rats had brains that were inherently "simpler" than those of the higher species. In an animal whose behavior was so limited, it could well be imagined that the brain might operate equipotentially. But in the higher species, especially in humans, the greater complexity of behavior indicated an increasingly functionally differentiated brain. Lashley dismissed this objection in the 1946 article, thereby managing to maintain the relevance of his theories to both rats and men.

So far, we have encountered nothing out of the antilocalizationist ordinary. Functions are not strictly localized in the variety of structures in the cortex, and this rule holds whether the brain is that of a rat or a human being. But then Lashley and Clark presented their second major objection to architectonic studies: that such cerebral maps take no account of the differences among brains of individuals of the same species. The goal of architectonics was to come up with a standard brain map for each species: a simple one for the rat, a more complex one for *Ateles,* an even more complex one for *Homo sapiens.* There was no room anywhere in this practice for differences in the brain maps of different monkeys; indeed, if it were the case that each monkey required its own individual brain map, then the usefulness of architectonics was thrown radically into question. Nothing,

[35] Ibid., 295.

in that case, could be learned from past brain maps, and every time the experimenter confronted a new specimen, he would be starting over.

It was precisely against this idea of the standard brain map that Lashley and Clark raised their objection. Lashley's interests lay in individual differences in ability; in his introduction to the 1932 book *Studies in the Dynamics of Behavior,* he speculated on the cause of differences between the ambitious and the slothful, the go-getter and the underachiever.[36] Later in his career, after the 1946 article, he continued to wonder and speculate about such differences. Differences between individuals concerned him much more than differences between species.

In the 1946 article, he and Clark affirmed that such individual differences must somehow be traceable to the differences in individual brain structure that they had observed – a belief that Lashley repeated to the end of his life. Emphasizing the pointlessness of the project of architectonics, Lashley and Clark declared that the " 'ideal' architectonic chart is nearly worthless because individual variation is too great to make the chart significant for a single specimen . . . ." And connecting these individual structural differences to differences in mental function: "The existence of such local variations in cell size, density and arrangement as we have found in different specimens of *Ateles* suggests a structural basis for individual differences in behavior." And finally, summarizing the point: "Marked local variations in cell size and density among individuals of the same species may constitute a basis for individual difference in behavior."[37]

With this second objection, we come to the crux of the matter. How is the second of Lashley and Clark's objections in accord with the first? Are they not, rather, in glaring contradiction? The first objection was that architectonic charts overemphasize and exaggerate the structure/function correlation; structural differentiation does not imply functional differentiation, and despite the brain's anatomical complexity, the major cortical areas function as equipotential wholes. Then came the second objection, that architectonic charts *under*emphasize the structure/function correlation; that different individuals have different brains, which means that they will behave and think differently. Lashley thus seems in this article to be speaking out of both sides of his mouth. On the one hand, he claimed, the architectonic maps require too much structure/function specificity; but on the other hand, they don't require enough. They are overly specific where they should be more vague, and too vague and general when what is needed is specificity. From the 1946 article, it seems clear that Lashley was objecting not to the architectonic effort to relate structure to function, but to a standard architectonic map that could apply to all individuals of a given species.

Does this mean that Lashley was trying quietly to jettison the idea of equipotentiality? Had it begun to outlive its usefulness? Was he simply engaging in his usual

[36] Lashley, ed., *Studies in the Dynamics of Behavior* (University of Chicago Press, 1936), vii.
[37] Lashley and Clark (1946), 298, 299, 300.

antilocalization rhetorical flourishes, while in the meantime discreetly taking on board elements of the localization theory?

The historian, of course, can only engage in educated speculation. My argument is that if we understand Lashley in his hereditarian context – a context that developed as his career progressed, but that in one form or another was always a part of his scientific work – we can begin to see how these two seemingly incongruous objections fit together. Lashley never gave up the idea of equipotentiality; for him it was a hereditarian idea, and his defense of it and attacks on localization in the 1946 article were as strong as ever. But equally a part of his hereditarianism was his biological determinist conviction that behavior, mind, intelligence, were products solely of the biology of the brain. As he had argued in 1923 that consciousness must be reduced to its physiological substrate, he now argued more than two decades later that differences in function must be traced to and interpreted in terms of differences in structure.

Both his argument for equipotentiality and his argument for structure/function correlation made sense within a hereditarian framework, and Lashley used both to support such a framework. Both were aimed at the same end; and for this reason I prefer to understand them in productive tension rather than in blatant contradiction. Equipotentiality emphasized the notion that the amount of functional brain mass could not be altered or improved during one's lifetime; meanwhile, the structure/function correlation explained why different individuals were different, an explanation completely dependent on inborn biology and not at all on environment and culture. Individuals were born with different brains; their different brains accounted for their different abilities; those brain structures, and hence their abilities, were unchangeable. In this way, Lashley's seemingly incongruous arguments in the 1946 article worked together to support his hereditarian convictions.

The 1946 article also shows how Lashley worked to defend equipotentiality and dismiss criticism, even in the face of mounting evidence in favor of localization. If he had wanted to give up equipotentiality, the mid 1940s would have been a good time to do so. Earlier in that decade, both American and British investigators had begun to map out the representation of the body that existed in the brain. When a particular part of the body was touched, which neurons in the brain received the sensation, and fired? If neurons in different cortical locations fired when different parts of the body were touched, could a "map" of the body then be discovered in the cortex? By 1941, many physiologists believed that such cortical maps did exist. For example, a group of researchers at Johns Hopkins including the physiologist Clinton Woolsey, working on cats and monkeys, revealed that the application of a brief tactile stimulus of low intensity to a very small cutaneous area evoked discrete surface positive potential waves in specific places on the contralateral cortex. These potentials, Woolsey and his coauthors argued, constituted the primary cortical response to sensory stimulation. Each region of maximal response was surrounded by a fringe or margin of submaximal response; when an adjacent part of the body was stimulated, a corresponding region in the cortex showed the maximal response.[38]

Similarly, the English physiologist E. D. Adrian reported, also in 1941, that he had drawn maps of the cortical "somatic receiving areas" of a rabbit, a cat, a dog, and a monkey. His experiments involved applying pressure to, say, the animal's forefoot, and measuring the location and intensity of the afferent electrical impulses reaching the cortex. Adrian, too, demonstrated that a cortical map corresponding to parts of the body could be drawn.[39] Clearly these reports gave more than a moment of triumph to the localizationists, and dealt Lashley's equipotentiality a blow from which it never fully recovered.

Lashley, however, deflected these criticisms, conceding as little as possible to this new localizationist upswing. It was true, he admitted, that the arrangement of the cortex preserved, at least to some extent, a faithful representation of bodily structure. For example, the striate cortex had a distinct structure because of its role in visual perception. This area of the cortex maintained the topography of the retina: as the retina sensed different parts of the visual field, the striate cortex represented the retina so as to preserve those geographical relationships. However, Lashley went on, other areas of the cortex had the same type of architectonic structure, yet their functions were quite different. The "somatic receiving area," the auditory cortex, and the gustatory cortex all had the same arrangement of cells as did the visual cortex, yet while the somatic and visual cortex preserved the spatial arrangement of the outside world, the auditory cortex had to preserve temporal, rather than spatial, sequences, and the gustatory cortex did not preserve any kind of organization whatever. Yet these areas had very similar architectonic structures.

In what sense, then, Lashley asked, was brain mapping a good way of studying brain function? Structure was evidently not a reliable index of function; what good, then, was a map of the structures that bore no relationship to the functions of the areas? Cortical maps proved useful for only two of the five senses; and they certainly provided no clue to how the higher functions of the brain (memory, reason, intellect) operated. If single ideas were not located in single brain cells, then what good was even a highly detailed map of the architecture of the cortex?

With the advantage of hindsight, we can see that for the most part, Lashley and his localizationist opponents were arguing past each other, with Woolsey and Adrian focusing on those aspects of brain function where localization held good, and Lashley focusing on those aspects where it did not. What is notable about this debate is how little each side seemed willing to concede to the other. Lashley

[38] W. H. Marshall, Clinton N. Woolsey, and Philip Bard, "Observations on Cortical Somatic Sensory Mechanisms of Cat and Monkey," *Journal of Neurophysiology* 4 (1941): 1–24.
[39] E. D. Adrian, "Afferent Discharges to the Cerebral Cortex from Peripheral Sense Organs," *Journal of Physiology* 100 (1941): 159–191.

ended his 1946 article by concluding that architectonic brain maps were little more than useless; he did not seem willing to budge an inch on equipotentiality (though to us it appears that he did make concessions to the localizers). Meanwhile, Woolsey, in a 1948 article on the structure and function of the limbic cortex, refused to concede Lashley's antilocalizationist point:

> Our results, therefore, are incompatible with the conclusions of Lashley and Clark ('46) who state that "the 'ideal' architectonic chart is nearly worthless . . . because the areal subdivisions are in large part meaningless, and because they are misleading as to the presumptive functional divisions of the cortex." In regard to the limbic cortex the criticism of Lashley and Clark is, we believe, without any substantial foundation. They appear to have greatly exaggerated some justifiable objections and they have unfortunately included the limbic cortex in their all too sweeping generalizations.[40]

Needless to say, this controversy was never resolved. But the fact that it was not can begin to give us a sense of the tremendous ideological significance that the theory of equipotentiality held for Lashley. He admitted that certain cortical areas maintained the topography of the retina or of the sensory surface of the body; or that structural differences in individual brains affected how those individuals functioned. Yet he did not interpret these seeming concessions to the localizationists as at all weakening the idea of equipotentiality. (Whether or not they actually did is another matter.) Despite these other findings, localization was still an outmoded theory for Lashley, and equipotentiality still provided "a cardinal touchstone for modernness of thinking."[41]

Lashley's lifelong support of equipotentiality, connected as it was to his hereditarianism, led him to argue against the improvability of mental function. But for Lashley, the debate over intelligence and its determinants did not end with his skirmishes with Dunlap and Pavlov. The notion that man was a creature built for progress also separated Lashley from his colleagues at the University of Chicago, especially from the neurologist C. Judson Herrick. Lashley's quarrel with Herrick, and with the tradition that Herrick represented, will occupy the next two chapters.

---

[40] Jerzy E. Rose and Clinton Woolsey, "Structure and Relations of Limbic Cortex and Anterior Thalamic Nuclei in Rabbit and Cat," *Journal of Comparative Neurology* 89 (1948): 325.

[41] Gardner Murphy, "Personal Impressions of Kurt Goldstein," in *The Reach of Mind: Essays in Memory of Kurt Goldstein,* ed. Marianne L. Simmel (New York: Springer, 1968), 34. Murphy's statement ran as follows: "Psychologists witnessed through the whole career of K. S. Lashley the renaissance of a belief in wholeness of central nervous function, as against the localization theory of the late nineteenth century, and for them the persistent problems of 'cerebral localization' made the Lashley emphasis a cardinal touchstone for modernness of thinking."

5

Psychobiology and Progressivism

The Rise of the Laboratory

In Lashley's neuropsychology, the laboratory, as opposed to the clinic or the field, was not simply the privileged site of knowledge production; it was the only source of reliable psychological fact. Lashley and his students cared little for the applied aspects of their discipline, not because they lacked interest in social control, but because for them the laboratory served as a substitute for society. Within its walls, they created a place where any social situation that was of concern could be simulated. By subsuming of all other aspects of life, the laboratory achieved preeminence in Lashley's neuropsychology.

The preeminence of the laboratory in Lashley's science had three main consequences for the production of psychological knowledge.[1] First, anything that could not be investigated in a laboratory was effectively stricken from the scientific record. What couldn't be studied within the laboratory walls wasn't science. Much of human psychology, consequently, was either excluded or redefined to fit inside a laboratory. Second, because experimentation with humans was strictly limited, the analogy between human beings and animals – particularly rats – took on a heightened significance and validity. Humans and the "lower" organisms were entirely comparable; any suggestion of a qualitative leap, of progress in evolution, was denied.

Third, the laboratory set limits to the kind of social conditions that could be reproduced and studied: in effect, it set limits to environmental variation. Moreover, because laboratory conditions were controllable and could be standardized, the environmental influences on behavior could be minimized. Environment was not necessarily dismissed, but it was marginalized. Here the divergence between Lashley's and Watson's approaches can be most clearly seen. Watson used his laboratory to demonstrate how completely behavior was produced by environment. For Lashley, however, the laboratory permitted an ever-sharper focus on the "natural" qualities of organisms: on the biological components of their psychol-

---

[1] For this analysis I have drawn on the essays in *The Rise of Experimentation in American Psychology*, edited by Jill G. Morawski (New Haven: Yale University Press, 1988), and on Kurt Danziger's approach in *Constructing the Subject* (New York: Cambridge University Press, 1990).

ogy, on their innate capacities. The laboratory permitted Lashley to emphasize those traits that would rise to the surface while the complex "social" factors were defined as inaccessible to scientific investigation.

In this way, the preeminence of the laboratory in Lashley's science helped to confirm his biological determinism and reinforce the hereditarianism that had been brewing in him all along. As his work evolved during the 1930s, 1940s and 1950s, Lashley looked to the science of genetics for the ultimate material cause of innate capacity, and his biological determinism became a full-fledged genetic determinism.

Genetic determinism was not, however, the inevitable concomitant of a laboratory-centered science of brain and behavior. At the University of Chicago, Lashley pursued his genetic determinist program in the midst of a group of scientists, laboratory psychobiologists of the so-called American school, whose work took a very different direction.[2] The crucial distinction between Lashley's school of neuropsychology and the American school of psychobiology lay in the stress each accorded the laboratory, especially with regard to society.

For Lashley and his students, the laboratory was preeminent, and social factors were considered only insofar as they could be accommodated within its walls. The psychobiologists of the American school reversed this trajectory: instead of bringing society into the laboratory, they took the results of their laboratory research out into the world. Those elements of human psychology that Lashley dismissed as unscientific – consciousness, moral value, spirituality – were for the American school the distinguishing characteristics of the human condition. For Lashley, two factors determined behavior: heredity and environment. He believed that standardization of the environment would cause hereditary capacity to bulk large, whether the subject was a human being or a rat. For the American school, on the other hand, there was yet a third determinant of behavior in man: free will. What mattered most to the American school psychobiologists were just those human characteristics that could not be captured in a laboratory, that set human beings apart from animals, and that helped to demonstrate mental, moral and cultural progress. This was not the conventional clash of two cultures, of Lashley's biological determinism versus the American school's humanism. As neuropsychologists and psychobiologists, the members of both the Lashleyan and the American schools were all biological determinists. It was the role of the laboratory that set them apart. The preeminent position of the laboratory in Lashley's research helped him to deny the existence of progress, while the laboratory's subordinate position in the American school's work helped its members to assert progress as an undeniable fact.

Together the American school and Lashley's school represent two major but

---

[2] On the meaning of the word "psychobiology," see Donald A. Dewsbury, "Psychobiology," *American Psychologist*, 46:3 (1991): 198–205.

little understood trends in American psychology and biology which I will call, respectively, "Progressive" and "anti-Progressive." The concept of Progressivism has long posed a problem for American historians, as it seems at once to include everything and mean nothing. Most historians can agree, however, that it was a broad-based movement for democratic social reform that lasted from about 1900 to 1917. Historians of science have made the concept relevant to the study of American science by showing how Progressive perspectives informed eugenics and behaviorism.[3] Here I will argue that Progressive precepts ran deep in American science, even among those scientists who explicitly disavowed both eugenics and behaviorism, and that those precepts continued to shape thought and practice in psychology and biology well beyond the conventional 1917 endpoint. I will try to clarify the meaning of Progressivism by emphasizing the root of the word – progress – and by contrasting it to a tradition in American neuropsychology to which it was most definitely opposed.

Of the two traditions, the Progressive American school of psychobiology was in its day the more broadly represented, attracting followers in neurology, embryology and anatomy, and of longer duration, flourishing from about 1890 to 1940. Lashley's anti-Progressive tradition had a smaller following of mainly comparative psychologists, active mostly during Lashley's career, from 1920 to the late 1950s. Yet it is Lashley's influence that continues to shape the present-day sciences of brain and behavior. While the American school scarcely has a follower today, the repercussions of the anti-Progressive tradition continue to be felt. (I will take up this argument about the Lashleyan influence in the Epilogue.)

Since the first major controversy of Lashley's mature career was with proponents of the American school of psychobiology, in this chapter I describe how they subordinated the laboratory in order to achieve their Progressive goals. In the following chapter I will examine the reaction of Lashley and his students to this Progressive tradition – their raising of the laboratory to a preeminent position and their development of a genetic determinist model of behavior.

Progress, Progressivism, and the American School of Psychobiology

During the past 25 years, American historians have gradually given up the notion of a coherent "Progressive movement" and instead argued that shifting and sometimes contradictory elements made up what has become known as "Progressive" thought. Daniel T. Rodgers, for example, has noted that several main ideals – antimonopolism, the use of the language of social efficiency, and an emphasis on

---

[3] See Donald Pickens, *Eugenics and the Progressives* (Nashville: Vanderbilt University Press, 1968); and John C. Burnham, "Psychiatry, Psychology and the Progressive Movement," *American Quarterly* 12 (1960): 457–465.

social bonds – were popular among different Progressive thinkers, for different reasons, and at different points in their history.[4]

At the same time, John Burnham has cogently argued that historians' difficulty in defining Progressivism stems from the narrowness of their criteria, and that they have missed the broad changes that Progressivism wrought in American society.[5] In the case of the American school of psychobiology, for example, Rodgers's supposedly common Progressive ideals lose much of their applicability. The psychobiologists were not businessmen and so did not speak or write about monopolies. They were openly opposed to scientific management, behaviorism, and other attempts at efficiency engineering in society. Though they did stress the importance of harmonious cooperation for the good of the whole, they also never lost their faith in the individual as the engine of social change.

Yet the psychobiologists of the American school can without doubt be classified as Progressives. They were clearly sympathetic to Progressive reforms: several prominent members of the American school, for example, were friends and helpers of Jane Addams, the Chicago social reformer and founder of Hull House. What joined the psychobiologists to the Progressive cause? What made Addams and the members of the American school all Progressives?

The distinguishing feature of the Progressive reformer was, I think, an unshakeable belief in social progress; and the Progressive psychobiologists contributed to the cause by demonstrating that progress was a fact of nature. The psychobiologists provided biological justification for the Progressive view that society was on a course continuing ever upward to a democratic utopia of peace and prosperity. The American school psychobiologists looked to nature as a source of guidance, and it told them just what the Progressives wanted to hear.[6]

The American school of psychobiology was distinguished by its vision of progress throughout the natural world. This "school" was actually a loose coalition of neurologists, anatomists, and embryologists, mainly at the University of Chicago and at Columbia, founded by C. Judson Herrick and his elder brother Clarence, and led during the first half of the century by the younger Herrick, Charles Manning Child, and George Ellett Coghill.[7] Important members included Oliver S. Strong, Arnold Gesell, Adolf Meyer, Frederick Tilney, Myrtle McGraw,

---

[4] Daniel T. Rodgers, "In Search of Progressivism," *Reviews in American History* 10 (1982): 113–132.

[5] John C. Burnham, "The Cultural Interpretation of the Progressive Movement," in *Paths Into American Culture: Psychology, Medicine and Morals* (Philadelphia: Temple University Press, 1988), 208–288.

[6] On a similar school in ecology, see Gregg Mitman, *The State of Nature: Ecology, Community and American Social Thought, 1900–1950* (Chicago: University of Chicago Press, 1992).

[7] For a discussion of the themes informing this school, see Sharon Kingsland, "Toward a Natural History of the Human Psyche: Charles Manning Child, Charles Judson Herrick and the Dynamic View of the Individual at the University of Chicago," in *The Expansion of American Biology*, eds. Keith R. Benson, Jane Maienschein, and Ronald Rainger (New Brunswick: Rutgers University Press, 1991), 195–230.

Elizabeth Crosby, C. U. Ariëns Kappers, John Black Johnston, Stephen Ranson, George Howard Parker, and Davenport Hooker. Many of them were associated with the *Journal of Comparative Neurology*, which the Herrick brothers founded in the 1890s.

Aside from the overarching vision of progress, four main beliefs linked these researchers and gave the American school its distinct program. First, the members of this school believed that structure and function formed an interdependent unity; one could not be studied without taking the other into account. One of their most popular avenues of research, for example, was to correlate changes in the structure of an organism with changes in the organism's behavior. Such a study could be pursued not only in *Planaria* or rats, but also in human beings, where the correlation became that between body and mind. The interaction and integration of body and mind in all aspects of life was an essential component of Progressive psychobiology, coming as it did out of the Baptist faith that inspired many members of the American school.

Second, the members of the American school believed that evolution was a process of adjustment to the environment, and that possession of a mind or consciousness helped the individual to adjust. This belief the psychobiologists borrowed directly from the pragmatist philosophers, especially John Dewey. An organism with a broader, more flexible repertoire of behavior could adapt more readily to a changing environment. Human beings adapted most successfully because their minds granted them a most complex range of behavior.

Third, for the psychobiologists of the American school, evolution meant progress; the terms were completely synonymous. "Evolution" referred not only to the process of phylogenetic development, but also to ontogenetic development; and it applied not only to these biological processes, but to changes in society and culture as well. All these types of evolution were equally progressive. Progress was obvious in the growth of the embryo, and in the evolution of the human species from its lower ancestors; social and cultural evolution were simply the continuations of these progressive trends.

Fourth, and last, the psychobiologists of the American school believed that evolution, particularly social and cultural evolution, could be controlled through education. Man, they believed, had an innate capacity to learn; by teaching people to live for the right values, a better, more harmonious, more peaceful, more prosperous society could be attained. The notion of controlling evolution usually implies eugenics, but the psychobiologists of the American school were not by and large eugenists. They did not believe enough in genetic determinism for that; and besides, they argued, improvement through selective breeding was much too slow a process. Education would bring results much more efficiently.

These four ideals can best be grasped not in the abstract but rather in their fulfillment in the work of the members of the American school. The work of the embryologist Charles Manning Child (1869–1955) shows how function and structure were integrated. From 1895 to 1937, Child was professor of embryology in the department of zoology at Chicago, where he developed his powerful and influential theory of gradients. Simply put, Child's theory stated that an organism developed through the creation within itself of different rates of metabolic activity. A mass of protoplasm remained undifferentiated until acted upon by a chemical or electrical stimulus. The region of the protoplasm most intensely affected by the stimulus responded by increasing its metabolic rate, displaying what Child called "a rise in the level of living."[8] In other less affected regions of the protoplasm the rate remained lower. This hierarchy of metabolic excitation Child called a gradient. The region of highest excitation became dominant and developed into the organism's most advanced structure, the head. Subordinate regions became lower structures, such as organs of digestion, reproduction, and locomotion.

The slope and position of these gradients were not at all predetermined by the structure of the organism; in fact, before the gradients appeared, the organism really had no specific structure. The gradient was determined only by the organism's response to a stimulus, in other words, by its behavior in response to an external force acting upon it. If the stimulus were constant and the response persisted, the resulting gradient became fixed as a structure. If the stimulus did not persist, it would produce only a transient physiological response and not lead to actual anatomical differentiation. Structure, therefore, was determined by behavior, by the organism's response to its environment. As Child wrote, "Apparently there is no escape from the conclusion that organismic pattern is not inherent in protoplasm, but arises in the final analysis from the relation of protoplasm to external factors."[9] And as he put it in *Physiological Foundations of Behavior,* the organism is "a behavior pattern in protoplasm."[10]

Child made it clear that this creation of structure and pattern from the organism's response to a stimulus was no different from learning and memory. In fact, learning and memory in the psychological sense were simply specialized aspects of the general developmental learning process. As the gradients arose in response to external stimuli and, in persisting, modified the further behavior of the organism, they were fundamentally similar to these higher psychological processes. Learning and memory were simply minute alterations in the structure of the nervous system from the impingement of and reaction to a stimulus, Child argued. Development was just a broadening of this concept to include all kinds of changes in structure resulting from behavior in response to a stimulus. Development, like learning, was a record in the protoplasm of behavior. "Viewed in this way the whole course of development is a process of physiological learning," Child wrote,

---

[8] Charles Manning Child, *The Origin and Development of the Nervous System from a Physiological Viewpoint* (Chicago: University of Chicago Press, 1921), 71.

[9] Ibid., 17.

[10] Child, *Physiological Foundations of Behavior* (New York: Henry Holt, 1924), 10.

beginning with the simple experience of differential exposure to an external factor, and undergoing one modification after another, as new experiences in the life of an organism or of its parts in relation to each other occur.[11]

There was no difference, therefore, between this general physiological memory and higher forms of memory in the nervous system. Man's capacity to learn, in which the members of the American school had so much faith, was for Child a general biological characteristic of all organisms.

Child believed that because education was an extension of growth, a basic biological process, man was therefore naturally fitted for it. This was a theme taken up by Chicago's social reformers. Among these was Jane Addams, the founder and director of Hull House, a settlement dedicated to improving the lives of slum dwellers. In addition to their direct work with the poor, Addams and other leaders in the settlement movement also pressed for legislation which resulted in, among other things, the establishment of juvenile courts.

The first juvenile court in the United States had, in fact, opened in Chicago in 1900, and the first "Psychopathic Institute" for the study of juvenile delinquency in 1909. In 1925, Addams held a conference to commemorate the founding of both of these institutions; the papers presented at this joint celebration were collected and published as *The Child, the Clinic and the Court,* and contributors included not only Addams herself, judges of juvenile courts, and directors of charities, but also Child and Herrick.[12] Both stressed their belief that learning from experience and ability to be educated were fundamental biological properties.

In his address, "The Individual and Environment from a Physiological Viewpoint," Child wrote that the form of an organism was dependent on its relationship to its environment; by changing these external conditions, the very structure and physiological relations of the organism could be altered. A head could be induced to develop in a region that normally gave rise to a tail, depending on the exposure of that region of the protoplasm to changes in the chemical and physical environment of the developing individual. Heredity and standardization of the environment helped to ensure that heads did not appear where tails should be; but Child felt his evidence convincing that environmental change could create deep structural modification.

In the human realm, this biological finding indicated that efforts at reform – changing the environment of the juvenile delinquent – could have profound effects on development. As the protoplasm could "learn" to develop differently, so could the juvenile offender. Child again stressed his idea that development of

---

[11] Ibid., 249.

[12] Addams et al., eds., *The Child, the Clinic and the Court* (New York: New Republic, 1927). Other speakers included the anthropologist Franz Boas; Herman Adler, the director of Chicago's Institute for Juvenile Research; and Julia Lathrop, the head of the Children's Bureau in Washington, D.C.

the embryo was continuous with the process of education: "The establishment of a persistent gradient is, in fact, the first step in the education of the protoplasm and the foundation of later behavior to the end of individual life."[13] Child used embryological development and education as interchangeable terms:

> Our experiments represent different educational methods and conditions and we find in development . . . that the same kind of protoplasm may give very different results according to its education.[14]

Child concluded his address by noting that he was not a "biological optimist," one who believed environment could turn anyone into anything; he did believe that heredity was an important limiting factor, but "such limitation affords no grounds for discouragement or inaction."[15] Child was therefore advocating a position that the social reformers found very compelling, grounding their political beliefs in what he felt was biological fact. For their part, the reformers were eager to hasten this reconciliation of politics and biology; they wanted to put their reforms on a scientific basis. This outlook was clearly reflected in the address by Augusta Fox Bronner, a pioneer in the study of juvenile delinquency:

> Scientific findings should become guides to treatment; they should find their greatest function, not in classifying and labeling, but in determining details of therapy.[16]

Progressive themes were evident too in the work of the other members of the American school. With Child and Herrick, the school's third founder was the embryologist George Ellett Coghill (1872–1941).[17] A protege of Clarence Herrick and close friend of C. Judson Herrick, Coghill spent his career investigating the relationship between structure and function in the nervous system of the salamander. He derived from his work a series of principles which, as both Coghill and C. Judson Herrick emphasized, were moral as well as biological: not only were they justified on a biological level, they also served as guides for human action. The principles provided not only knowledge but also wisdom, as Herrick put it.

Coghill had begun his study of "the nervous system as an approach to psychology and philosophy" on the inspiration of Clarence Herrick. The elder Herrick

---

[13] Child, "The Individual and the Environment from a Physiological Viewpoint," in *The Child, the Clinic and the Court,* 150.

[14] Ibid., 151.

[15] Ibid., 155.

[16] Augusta Fox Bronner, "The Contribution of Science to a Program for Treatment of Juvenile Delinquency," in *The Child, the Clinic and the Court,* 79. Bronner was codirector, with William Healy, of the Judge Baker Guidance Center of Boston.

[17] On Coghill, see Hamilton Cravens, "Behaviorism Revisited: Developmental Science, the Maturation Theory, and the Biological Basis of the Human Mind, 1920–1950s," in *The Expansion of American Biology,* eds. Keith R. Benson, Jane Maienschein, and Ronald Rainger (New Brunswick: Rutgers University Press, 1991), 133–163.

offered him his microscope and laboratory space, and seemed to have an un-bounded faith in the young man's capacity to learn. Coghill subsequently received a Ph.D. in biology at Brown University and studied with Theodore Boveri at Würzburg; he taught at Pacific University and Willamette University in Oregon, then at the University of Kansas, and ended his career at the Wistar Institute of Anatomy and Biology in Philadelphia. C. Judson Herrick wrote his biography in 1949, in which he presented Coghill as a "representative man of science" and as an example to the young because he was both naturalist and philosopher, living the psychobiological ideal by combining scientific research with a humanistic concern for the moral value of that research.[18]

The first and most important of Coghill's principles was that the organism developed as an integrated whole: that its behavior and its structure were corre-lated during its development. Changes in behavior and changes in structure oc-curred together. The organism did not develop fully and then begin to act, nor did its various behaviors appear as discrete reflexes which only later were coordi-nated. The whole organism was active from the start, and this activity grew along with structural development.

The second of the Coghillian principles was that preneural and neural activity were continuous: there was no sharp disjunction where unconscious activity ended and conscious mentation began. There was a continuity between behavior and mind. As Herrick imagined Coghill saying in a philosophical dialogue, "men-tation . . . is a total-pattern type of activity common to all organisms and . . . there has been a progressive expansion and individuation of this pattern in evolutionary history and personal development."[19]

Taken together, these two principles formed the justification for psychobiology. Because mind was simply a type of activity or function, and because activity and structure were closely coordinated throughout development, the study of one had to involve the other; the mind could not be investigated without an understanding of the body, nor the body without an understanding of the mind. The American school's program of psychobiology, therefore, which presented the mind-body complex as an interdependent unity, found justification in Coghill's study of salamander biology.

The idea that structure and function were coordinated during development led to the third of Coghill's principles. Neither structure nor function was prior in development; rather, the organism possessed an "intrinsic motivation" which preceded both its behavior and its structure. Members of the American school connected this notion of intrinsic motivation with their belief that man possessed free will: they translated this psychological capacity into biological terms. The

---

[18] Charles Judson Herrick, *George Ellett Coghill, Naturalist and Philosopher* (Chicago: University of Chicago Press, 1949), 17, v.
[19] Ibid., 216.

emphasis on intrinsic spontaneity was at the heart of the American school and the faith of its members in social reform. It supported the belief that the organism was not merely buffeted about by its heredity or its environment, but contained the power within itself to control its own course. Members of the American school emphasized this intrinsic activity by calling their approach to psychobiology "dynamic." This is what Coghill meant when he said

> I would emphasize the fact that the nervous system is more than a stimulus-response mechanism. There are intrinsic sources of motivity. The individual first acts on its environment and then more or less reacts to its environment. Accordingly, in our present knowledge of the correlation of structure and function in the development of behavior, there is no ground for a scientific determinism or fatalism.[20]

C. Judson Herrick himself expanded on this idea: self-determined motivity, he wrote, "emancipates man from inevitable bondage to heredity and environment and endows him with some measure of purposive control both of these and of his own destiny."[21]

The central importance of Coghill's work was its correlation of the biological and the psychological. Coghill demonstrated that human mental capacity – ability to learn (intelligence), freedom of will, capacity to envision the future – were literally properties of protoplasm. Because these psychological capacities were firmly grounded in biology, there was biological justification for social reform, which depended on man's ability to learn and thereby to control his own destiny.

Three examples will show that this justification was quite literal. First, Coghill did not merely make a vague equation between learning and growth. He had actually found evidence that nerve cells did not cease to grow once they became conductors of electrical potential; rather, they continued to grow by microscopic increments "in strictly embryonic fashion."[22] They were never simply static conductors, like the lines of a telephone switchboard; they always retained their dynamic character. This neural growth was the equivalent of learning because it resulted in an expansion of the organism's behavioral repertoire.

There were clear connections between this principle and Child's idea that development was protoplasmic learning. The equivalence between such a basic biological process as growth, and such an apparently advanced psychological process as learning, showed that the capacity to learn inhered in our cells, at the most basic biological level. Coghill's treatment showed how the biological and the psychological could not be separated. A change in structure through growth carried with it changes in function; the capacity to change in response to the

---

[20] Quoted by Herrick in *Coghill*, 137.
[21] Ibid., 138.
[22] George Ellett Coghill, *Anatomy and the Problem of Behaviour* (Cambridge: Cambridge University Press, 1929), 81.

changing requirements of the environment Coghill defined as intelligence. Intelligence, capacity to learn, was therefore the same as the biological capacity to grow.

Second, Coghill had found that behavior developed as an integrated whole, but that specific reflexes became individuated against this integrated background. He called this the Gestalt conception of behavior, borrowing the term from a school of psychology which had used it to refer to the perception of a figure against a ground. Again Coghill demonstrated the continuity between an apparently "low" form of activity and the higher psychological processes. The implication was that human perception was merely the extreme end of a continuous process of behavioral development.

Third, Coghill believed that "neural overgrowth" provided biological justification for the idea that man can control the future. Nervous organization that developed before it was actually functional represented the "potentialities of behavior" that could come to full expression only in the future.[23] For Coghill, this was the "mechanistic equivalent" of man's ability to plan ahead. The fact that the salamander's brain possessed advanced optic centers before the optic nerve approached the brain was the biological equivalent of planning for the future. Such neural overgrowth occurred during both ontogeny and phylogeny, since Coghill assumed that the two processes were fundamentally similar.

Coghill advocated such parallels between biological and psychological processes because he wanted to show that mind and body were not two separate entities. Yet even as he demonstrated the biological basis of the human mind, he did not deprive the mind of its power. The fundamental similarity between free will and an embryo's "internal motivity" did not mean to Coghill that free will was therefore an illusion. Quite the contrary: free will was the natural extension of this fundamental biological process, just as the ability to plan ahead was the extension of neural overgrowth.

Herrick held up Coghill as an example to the young because Coghill had managed to include in his research a concern for its moral worth. Coghill had shown that the best scientific work had a strong humanistic component. Biological fact and social value could not be separated; moral truths were to be found in nature. Politicians and social reformers who tried to conform man's nature to unnatural systems like socialism and fascism were bound to fail. Likewise, biologists who ignored the humanistic value of their work rendered themselves obsolete and ineffectual: they had nothing to say to the world. As Herrick wrote at the end of his Coghill biography, "Scientific facts are not worth what it costs to discover them unless they can be so interpreted as to lead to value-judgments and as guides to more satisfying purposeful action."[24] Ultimately this concern with the

---

[23] Ibid., 92–4. The neural overgrowth was not nonfunctional (function and structure were always correlated); rather, its primary purpose could not yet be fulfilled.

[24] Herrick, *Coghill*, 229.

humanistic significance of science also had its roots in the American school's psychobiological ideal: as the realms of mind and body were completely integrated, so biological fact and social value could not be separated.[25]

The equivalence of growth and learning was not a vague theory divorced from real life. It was applied directly to the classroom by the psychologist Arnold Gesell (1880–1961), director of Yale's Clinic of Child Development. Gesell used Coghill's biological principles to justify certain methods of educating children. He was unstinting in his praise of Coghill's work: "[Coghill's] three lectures on 'Anatomy and the Problem of Behavior' bid fair to become classic. The thin volume which contains them belongs on a five foot shelf."[26]

By 1940, the Second World War had begun and Gesell, echoing national themes, had become alarmed that democracy was in peril and that improper education was the cause. In his address to the Maine Teachers' Mental Hygiene Association, Gesell identified his concern with several larger contemporary developments. President Roosevelt had called a White House Conference on Children in a Democracy. The American Federation of Teachers had taken as its theme "Equality of Educational Opportunity to Save Democracy." The General Board of Education had launched a program to determine whether schools were "in accordance with democratic philosophy."[27] The Educational Policies Commission had recommended the fostering of democratic ideals through education.

Gesell believed that to preserve democracy required democratic teacher-student relationships, which could be established only if the teacher understood the laws of growth. "[G]rowth is the key concept for a sound philosophy of education," Gesell wrote, "and for the mental hygiene of the teacher-pupil relationship in a democracy."[28] Teachers had to realize that their students underwent a developmental process that shaped them as individuals: at each stage they should be appreciated and understood; they should not be expected to be exactly alike or to conform to some unrealistic ideal.

Coghill had emphasized that growth was the organic creative process: when the organism ceased to grow, it ceased to be intelligent. For Gesell this meant that children should be expected to go through stages. As Child had stressed the development of individuality, for Gesell each child was "an individual, with inborn propensities, with inherent constitutional characteristics."[29] Teachers had to recognize that growth was a necessary part of individuality: "The task of child

[25] Adolf Meyer also argued for the joining of fact and value against John B. Watson. See Ruth Leys, "Meyer, Watson and the Dangers of Behaviorism," *Journal of the History of the Behavioral Sciences* 20 (1984): 128–151.

[26] Arnold Gesell, "Scientific Approaches to the Human Mind," *Science* 88 (1938): 227.

[27] Gesell, "The Teacher-Pupil Relationship in a Democracy," *School and Society* 51 (1940): 194.

[28] Ibid., 198.

[29] Ibid.

care is not to mould the child behavioristically to some predetermined image, but to assist him step by step, guiding his growth."[30]

Gesell was not alone in applying the biological themes of the American School directly to the classroom. The neurologist H. H. Donaldson wrote on neurology and education, as did Frederick Tilney, the Columbia neurologist.[31] And the combination was evident in the lives as well as the works of the American school members. Elizabeth Crosby, Herrick's protegee, was a high school principal before taking a position as neuroanatomist at the University of Michigan. John Black Johnston, Lashley's teacher of neurology at West Virginia University, turned to college administration at the University of Minnesota and in 1930 wrote a book called *The Liberal College in a Changing Society*.[32] This work exemplified one of the American school's main themes: control of evolution toward democracy by education. As Johnston warned, "Human evolution must go on either by starts and jerks, by quakes and violence or by orderly processes directed by knowledge. This knowledge can be secured only through higher education."[33]

---

[30] Ibid.

[31] Donaldson, *Growth of the Brain: A Study of the Nervous System in Relation to Education* (New York: C. Scribner's, 1895); Tilney, *The Master of Destiny: A Biography of the Brain* (Garden City: Doubleday, Doran and Co., 1930).

[32] J. B. Johnston, *The Liberal College in a Changing Society* (New York: The Century Co., 1930).

[33] Johnston, 41. Because of its connections to education and child guidance, the American school of psychobiology made room in its program for "women's work." Indeed, the school had a number of prominent women members. One of these, the psychologist Myrtle McGraw (1899–1988), exemplifies the American school's belief in the inseparability of science and its social implications. During the first half of her career McGraw studied the behavioral development of children; in the second half she used her science to teach women to be better mothers. Like other members of the American school, McGraw was directly inspired by the writings of John Dewey, even writing to ask if she might study with him. Dewey, in turn, made it possible for her to attend Teachers College, Columbia University, by engaging her to type the manuscript of his book *Art as Experience* (1934).

McGraw obtained a master's degree in religious education and spent a year in Puerto Rico attempting missionary work – for which, she later wrote, she was not fit. As for many other members of the American school, for McGraw the alternative to religion was science. She returned to Columbia and worked toward a Ph.D. in child psychology under Helen Thompson Woolley, who had herself been a student of Dewey's. McGraw's dissertation analyzed the performance of white and black infants on a standardized infant test; she reported finding a " 'slight but consistent superiority' of the white students over the Negroes" in both physical and mental traits. (See "A Comparative Study of a Group of Southern White and Negro Infants," *Genetic Psychology Monographs* 10 [1931]: 94.)

In 1930 she began work with Frederick Tilney, the director of the Neurological Institute of Columbia's College of Physicians and Surgeons, and herself became associate director of the Normal Child Development Study at Babies Hospital of Columbia Presbyterian. There she did her famous work on the swimming reflex in infants. This tenure lasted twelve years, until 1942, when her grant was discontinued. By then she had begun what she called her "mother period": she had married "a brilliant scientist-engineer" several years earlier, and spent the 1940s bringing up their children, and doing some occasional teaching. (See Lewis P. Lipsitt, "Myrtle B. McGraw [1899–1988]," *American Psychologist* 45 [1990]: 977.)

In 1957 McGraw received an offer from Briarcliff College, a women's college near her home in Hastings-on-Hudson, to set up a laboratory like the one at Babies Hospital. The purpose was to have the students themselves handle the babies. McGraw believed that simply wanting children was not

The stress on evolution in the work of the American school was accompanied by an emphasis on embryology. In fact, the biologists of this school considered embryology and evolution to be parallel, progressive processes. As Coghill had traced the development of behavior and structure during the morphogenesis of the salamander, Frederick Tilney traced brain development and expansion of mental capacity through the evolution of the human race. Both Coghill and Tilney believed that past progress was indicative of future progress, and the point of their science was to demonstrate that control of evolution could bring about such progress. Coghill wrote, "man is, indeed, a mechanism, but he is a mechanism which, within his limitations of life, sensitivity and growth, is creating and operating himself."[34] Tilney saw the same opportunity for control of the future through control of the evolutionary process.

An M.D. as well as a Ph.D., Tilney spent his career at Columbia studying brain development during the growth of the individual as well as throughout the evolution of the species. Like the other members of the American school, he also worked actively for the betterment of society. It was said of him that

> he was constantly in the arena, battling for civic betterment and dealing with the daily adventure of the practice of medicine. . . . He was to be found on Municipal Committees for the better cleaning of streets, for the more decent conduct of our city traffic, and he was in the forefront of every battle for higher standards in education.[35]

Tilney envisioned a peaceful society based on a scientific knowledge of the development of the brain. In an article on "Behavior in Its Relation to the Development of the Brain," Tilney wrote that the ideal societies of Plato, St. Augustine, and H. G. Wells could be achieved, but only through a knowledge of neurology.[36] In his popular book *The Master of Destiny* (1930), Tilney explained how the development of "a better brain" was possible by means already well within our grasp – namely, eugenics and education. Progress was inevitable if these two were applied correctly, Tilney argued; our understanding of the evolu-

enough: teenagers had to know how to be caretakers themselves before they started bearing their own children. She was essentially providing young women with a knowledge of motherhood based on a scientific study of infancy and development; her former students wrote to her in gratitude "after they had become mothers themselves." McGraw, therefore, exemplified the ideals of the American school by doing scientific research that addressed the problems of society. (See Myrtle B. McGraw, "Memories, Deliberate Recall and Speculations," *American Psychologist* 45 [1990]: 937.)

On child psychology as "women's work," see Margaret Rossiter, *Women Scientists in America: Struggles and Strategies to 1940* (Baltimore: Johns Hopkins University Press, 1982), 203–4, 245.

[34] Quoted in Herrick, *Coghill,* 222.

[35] Foster Kennedy, "Frederick Tilney, M.D.: An Appreciation," *Journal of Nervous and Mental Diseases* 89 (1939): 267.

[36] Frederick Tilney and Lawrence S. Kubie, "Behavior in Its Relation to the Development of the Brain," Part I, *Bulletin of the Neurological Institute of New York* 1 (1931): 229–313; Part II, *Bulletin of the Neurological Institute of New York* 3 (1933): 252–358.

tionary process allowed us to control it and turn it to our advantage: "The embarrassments of the laggard fraction of humanity would thus be overcome."[37]

In addition, Tilney's neurological research had demonstrated that during the course of evolution, the human brain had noticeably increased in size, a sure indication of past progress. This size increase Tilney believed was brought about by use, and he found the same correlation between use and structure during individual growth in his study of brain lipoids. These lipoids, providing insulation for the brain's tracts and pathways, were highly developed in the adult, and most highly in the most intelligent adults.[38] Not only did this mean that brain structure was therefore a reliable index of intelligence, but it also showed that extended use led to improved structure which led to increased levels of intelligence. The achievement of progress was then very much within our own power. In addition to eugenics, it was education that would allow man to become the master of destiny:

> Greater than the power of armies, more compelling than the military force of the entire globe, is the peaceful sway which education may exert in the satisfactory reshaping of existence.[39]

Herrick and the Limits of the Laboratory

To this point, I have portrayed the psychobiologists of the American school as thoroughgoing biological determinists, freely applying the results of their laboratory studies in the world and seeing everywhere in society their biological principles fulfilled. Child, for example, often made the analogy between the growth of organisms and the development of democratic societies, arguing that as democracies were the highest form of human organization, so organisms that functioned on "democratic" principles were likewise the most advanced forms of life. There were, however, limits to the biological determinism of the American school, limits that were most clearly articulated by C. Judson Herrick. Throughout his career, Herrick was faced with the difficult task of maintaining a delicate balance between biological determinism, on the one hand, and on the other the freedom of choice he believed to be the distinguishing mark of the human condition.

Herrick's attempt to achieve this balance is most evident in his writings on consciousness. He called himself a "radical mechanist," by which he meant that he saw all events as having specifiable causes. There was no room for mystical vital forces in Herrick's world: it was a deterministic, natural world, terms he used interchangeably with "mechanistic." Herrick explained all aspects of the world in terms of natural causes and effects, from the development of a salamander embryo

---

[37] Tilney, *The Master of Destiny,* 334.
[38] Tilney and Joshua Rosett, "Brain Lipoids as an Index of Brain Development," *Bulletin of the Neurological Institute of New York* 1 (1931): 28–71.
[39] Tilney, *Master of Destiny,* 340.

to the capacities of the human mind. Consciousness was a thoroughly biological phenomenon, explicable only through an understanding of the cerebral cortex. Yet in this mechanistic, deterministic scheme, Herrick found a place for free will, without compromising the naturalness of his world. To understand how this seeming contradiction made perfect sense to Herrick is to grasp the motivation for his life's work.

Herrick's explanation was simple: consciousness was a clearly observable fact in human life. People had ideas, and these ideas obviously determined their subsequent actions, and their actions in turn influenced their thoughts. Consciousness was not completely separate from biology, as the parallelists would argue, nor could it be dismissed in favor of the exclusive study of the body, as the behaviorists maintained. Both alternatives were equally unscientific: one put mind in a metaphysical realm beyond investigation, the other plainly ignored the evidence. Both missed the central fact of life: the interaction of mind and body in all behavior. Consciousness had definite consequences in the material world, and was therefore a natural event.

Once mind was accepted as a natural fact, Herrick argued, control of conduct became possible. This principle was central to his Progressive outlook. Free will was not incompatible with determinism; rather, because it was a product of the body, it was fully natural. As part of nature, it could interact with matter. A mind that floated off in some metaphysical realm had no issue among the objects of experience; but a mind that was entirely a product of the biology of the brain could have a definite influence in the material world. Exercising their free will by making rational choices, people could change their environments. Herrick's psychobiology gave the optimistic message that change was possible: people were able to make choices, and if they made up their minds to make the right choices, they could have real effect on all aspects of their existence. There would be no miracles, Herrick warned, since all change had to happen through deterministic processes, but there was hope. He wrote, "The acceptance of these plain facts of experience without reserve or any metaphysical distortion gives us a practicable approach to all the big problems of human life and experience with the technique of natural science."[40]

Herrick delivered his psychobiological message to all kinds of audiences. To philosophers he argued not that Alfred North Whitehead's conception of "organism" should replace mechanism, but rather that organism should be recognized as a special case of mechanism. As long as they operated by "lawfully ordered cause and effect sequences" all objects, whether artificial or natural, were machines – including muscles, glands, reflex arcs, and the human cerebral cortex.[41] Herrick's mechanistic perspective denied the existence of supernatural agencies, while al-

---

[40] Herrick, "Mechanism and Organism," *Journal of Philosophy* 26 (1929): 596–7.
[41] Ibid., 593.

lowing machines not only to be controlled by their environments, but also actively to control them. His insistence on the term "mechanism" was important because it strengthened his defense against those who, with Lashley, would charge him with mysticism.

At the Juvenile Court commemoration in 1925, Herrick told the audience of social workers, biologists, anthropologists, and sociologists that consciousness is a product of a deterministic world and can interact mechanistically with it.[42] But consciousness also implied freedom of choice and therefore an ability to improve one's situation. For Herrick this reasoning was perfectly consistent: the will as a natural force could be the cause of events in the physical world, like a stone rolling down a hill. But the will gave human beings the power to change the course of events if they chose, and therein lay the hope for social progress. The key to the process lay in education: people must be educated to work for a democracy founded on harmonious cooperation and pacifism, in which everyone achieved personal profit.

Herrick believed that human beings had, through the evolution of consciousness, passed beyond the biological realm and reached a higher psychological level. Biological laws were still valid on this level; yet we were given conscious control over our own evolution that lower animals did not possess. The dimension of mind had been added, and mindful human beings had a new power of will. Biological evolution had brought us this far, but now it was up to our own free will to direct the course of psychological, social, and cultural evolution. Progress would continue, but only if human beings were educated, lived in harmony with the laws of biology, and made the right choices. Because he was not merely a biological creature, man was in charge of the future:

> My judgment of the probable course of future events and my forecast of the probable consequences of my own proposed action are determining factors in shaping a decision or "making up my mind" to a purposeful choice just as truly as are my hereditary dispositions, my established habits, and my temporary physiological condition of hunger, fatigue, depression or exuberance of vital tone – and they may be by far the most significant factors.[43]

Consciousness was for Herrick beyond the control of both heredity and environment and in that sense was free; but as a product of the body it was completely within the realm of the natural.

---

[42] Herrick, "Self Control and Social Control" in *The Child, the Clinic and the Court,* eds. Jane Addams et al. (New York: New Republic, Inc., 1927), 156–177.

[43] Herrick, *Brains of Rats and Men: A Survey of the Origin and Biological Significance of the Cerebral Cortex* (Chicago: University of Chicago Press, 1926), 337.

Herrick's optimism that progress in the psychological, social, and ethical realms was bound to occur if we set our minds to it was based on his belief that biological evolution had followed a progressive course. "In organisms we have not merely the perpetuation of specific patterns of structure and behavior throughout metabolism, growth and reproduction as formulated in our laws of anatomy, physiology, embryology and heredity, but also progressive change in these patterns – evolution."[44] He constantly stressed the differences between man and the lower animals rather than the closeness of their relationship. In discussing the evolution of the learning process, for example, Herrick wrote: "The tremendous enlargement and complication of this mechanism as we pass from the highest living brutes to the lowest surviving races of men is indicative of a gap in the phylogenetic series of wide extent . . ."[45] Embryological development, which Herrick believed was parallel to evolution, also helped him to emphasize progress: progressive complexity was undeniable in the growth of an embryo to an adult. Herrick's collaboration with Child on the development of physiological pattern, and with Coghill on the differentiation of the nervous system, helped him to demonstrate the similarities between embryology and evolution, and to underscore the progressive nature of both processes.

Herrick's emphasis on progress affected both the content and the practice of his science. He argued strongly that biological determinism should have its limits. Free will might be the psychological analogue to "intrinsic motivity," but it was also infinitely more powerful than intrinsic motivity; it brought about effects of a radically different order. Herrick never forgot that all psychology depended upon its biological substrate and that society could never escape its biological foundations. Concern with the moral value of biological facts was necessary, he believed, if we are to perceive the "natural" tendencies of society. Yet while he stressed the biological basis of the human personality, he emancipated it from biological bondage; he granted human free choice the power to rise above instinctive and habitual behavior patterns and set society on a different course.

Similarly, Herrick's Progressivism also affected his scientific practice: he saw definite limits to laboratory experimentation. The dimensions of the human personality, and the psychological level at which they emerged, could not be adequately captured in a laboratory. They required for their full expression a complex society presenting a broad range of choices and possible outcomes, and such conditions were impossible to replicate within a laboratory. Human psychology could never fulfill its potential in the laboratory's controlled environment. Society was by definition the web of relations outside the laboratory; if it were simplified and made amenable to laboratory study, it would cease being society.

[44] Ibid., 285.
[45] Ibid., 220.

Herrick's recognition of the inability of biological determinism and of laboratory experimentation to do justice to human psychology and society sprang directly from his belief in progress through emergent evolution.[46] This recognition is the crucial difference between the Progressive and the anti-Progressive traditions in psychobiology. While Herrick believed that the laboratory was not and could never be the complete answer to society's problems, his younger colleague Lashley thought that the laboratory experiment was the only way human psychology was ever going to be understood. How Lashley and his school came to develop this perspective I will discuss in the next chapter.

---

[46] A belief in "emergent evolution" – that complex processes like "life" and "mind" emerge at increasingly complex levels of biological organization – characterizes many of the important biologists of the first half of the twentieth century, among them C. Lloyd Morgan and W. M. Wheeler. See Sharon Kingsland's discussion of H. S. Jennings's views on emergent evolution in comparison with those of Herrick in "Toward a Natural History of the Human Psyche," especially pages 217–220.

6

Psychobiology and Its Discontents:
The Lashley-Herrick Debate

I have found discussion with most of the radical behaviorists futile. We seem to speak different languages – common words do not mean the same to us . . .

Charles Judson Herrick[1]

Rats are not men. . . . Men are bigger and better than rats.

Charles Judson Herrick[2]

When Lashley arrived at the University of Chicago in 1929, he associated himself with a group of scientists whose work was permeated with the hope for human betterment. He had been recruited to the department of psychology at Chicago in order to help advance the program in "psycho-neurology," the integration of psychology and biology envisioned by Herrick and his colleagues.[3] While Lashley supported the American school's goal of unifying the sciences of brain and behavior, he became deeply opposed to their Progressive social ideals. He adopted their general embryological perspective, and Child's gradient theory in particular, but consistently rejected the implications of progress that were supposed to go along with it.

Lashley and Embryology

Lashley's use of the embryological model in his own work correlating brain function and behavior did not spring up instantaneously upon his arrival in Chicago. As early as 1917 he had been self-consciously borrowing terms from that field. References to Child's gradient theory, however, began to appear in Lashley's work only in 1926 and continued until 1935, when Lashley left Chicago for Harvard.

---

[1] Herrick to Adolf Meyer, no date [probably late 1920s], Adolf Meyer Archive, Series I, Unit 1674, Folder 2, Alan Mason Chesney Medical Archive, The Johns Hopkins Medical Institutions.

[2] Herrick, *Brains of Rats and Men*, 347, 365.

[3] On Chicago psycho-neurology, see Bonnie Ellen Blustein, "Percival Bailey and Neurology at the University of Chicago, 1928–1939," *Bulletin of the History of Medicine* 66 (1984): 90–113. Also Blustein, "Medicine as Biology: Neuropsychiatry at the University of Chicago, 1928–1939," *Perspectives on Science* 1 (1993): 416–444.

The analogy grew from his first use in 1917 of the term "equipotential." He applied this term to the brain, while embryologists, in particular Hans Driesch, had applied it to the embryo.[4] Lashley later credited Driesch with the introduction of the word.[5] But Lashley's adoption of the word, and his pursuit of the analogy, along with Herrick's and Child's, came from the desire of these "radical mechanists" to explain in nonvitalistic terms the phenomena that Driesch had observed.

Driesch had used the idea of a nonmaterial "entelechy" to explain restitution of structure and function after damage: that is, how half an embryo could generate a whole adult. Driesch theorized that the vital force intervened and redirected the morphogenetic process along a normal course. He inspired an almost universal negative response: some biologists questioned the validity of his results; others accepted his observations but dismissed his vitalistic explanation as unscientific.

For this latter group, the problem remained how to deal with his results. A common strategy was to use Driesch's own evidence and even his language without any vitalistic overtones. For example, in 1913 Driesch published *The Problem of Individuality;* two years later Child came out with *Individuality in Organisms,* which dealt with the same problems of differentiation and maturation of structure and function but avoided vitalism. Likewise, in their monumental two-volume comparative neurology, Kappers, Huber, and Crosby used Driesch's word "entelechic" to describe the brain's developmental stages but added the disclaimer: "The expression 'entelechy' is not used here as a property apart from the organism, superimposed upon it, but as an inherent organic property."[6]

Similarly, Lashley, who had received from his Ph.D. adviser H. S. Jennings "a violent aversion to vitalism," used "equipotentiality" to acknowledge Driesch's observation but negate his interpretation.[7] Lashley's reliance on embryological work to illustrate psychological principles implied that embryology should be a model for psychology in method as well as in substance: just as the embryologists had explained Driesch's results without his vitalism, the psychologists could explain the phenomenon of mind without recourse to nonphysical forces.

Lashley's first extended reference to Child's work appeared in his 1926 article on the relation between cerebral mass, learning, and retention. At the end of the paper, Lashley noted the common problems of neural function and of morphogenesis:

[4] Lashley and Franz, "The Effects of Cerebral Destruction Upon Habit-Formation and Retention in the Albino Rat," *Psychobiology* 1 (1917): 71–139.

[5] Quoted in Fred A. Mettler, "Cerebral Function and Cortical Localization," *Journal of General Psychology* 13 (1935): 397.

[6] C.U. Ariëns Kappers, G. Carl Huber, and Elizabeth Caroline Crosby, *The Comparative Anatomy of the Nervous System of Vertebrates, Including Man* (New York: MacMillan, 1936) xiin.

[7] Lashley's phrase, a "violent aversion to vitalism," is quoted by Darryl Bruce, "Lashley's Shift from Bacteriology to Neuropsychology, 1910–1917, and the Influence of Jennings, Watson and Franz," *Journal of the History of the Behavioral Sciences* 22 (1986): 30.

In both cases it is the pattern or total relationship of parts as well as particulate structures which determines the final product. The best fitting theory of the mechanism of development is that which appeals to physiological stresses, as elaborated by Child ('23) in his discussion of physiological gradients, and it seems probable that further development of theories of neural function must proceed along somewhat similar lines.[8]

Three years later, in *Brain Mechanisms and Intelligence,* Lashley developed a theory of neural function based on Child's theory of physiological gradients. This was his most detailed exploration of the analogy, but it is remarkable for its tentative character: he introduced it at the very end of his book, in a few pages, and stressed that it was speculative.[9] According to Lashley, the gradient systems that coordinated development did not cease to function in the mature organism, but continued to mediate its activities. When the organism was confronted by two stimuli of different intensities, Lashley theorized, two points of excitation were established at random in the neurological fields of its cortex. These two points were differentially excited: the more highly excited one corresponded to the more intense stimulus, the other to the less intense stimulus. Between these two differentially excited points was established a gradient of electrical action or chemical diffusion; that is, the two points were not isolated from each other, but set up a relationship with each other. Similar relationships, or gradients, were also established between these two points and others further afield in the cortex.

The advantage of hypothesizing such gradients for Lashley was that they required no localization of function in the cortex. They therefore could explain his principles of equipotentiality and mass action. The same gradient could be established anywhere in the thin wide sheets of gray matter in the cortex, as long as some two points were excited in the same way and in the same relationship. It was the relationship, not the specific location in the brain, that mattered. Once the gradient was developed, a series of motor acts could be initiated. Again, the point was that large sections of the brain could be destroyed, and yet the same ratios of excitation could be established and the same motor acts carried out, because the requisite gradient was independent of any specific locus.

The analogy answered Lashley's need for a mechanism to explain his observations of equipotentiality and mass action, yet he was hesitant to commit himself to it wholeheartedly. "I have not suggested this hypothesis as a picture of the actual processes in the central nervous system," he concluded, "but only as an illustra-

---

[8] Lashley, "Studies of Cerebral Function in Learning VII: The Relation Between Cerebral Mass, Learning and Retention," *Journal of Comparative Neurology* 41 (1926): 45–6.
[9] Lashley, *Brain Mechanisms and Intelligence: A Quantitative Study of Injuries to the Brain* (Chicago: University of Chicago Press, 1929), 166–172.

tion of the direction in which the facts of neural plasticity seem to force our speculations."[10]

That same year, 1929, in his presidential address to the American Psychological Association, Lashley again emphasized the analogy between his findings in neurology and Child's embryological work. Theorizing that he could slice off the cerebral cortex, turn it around and replace it, and that the animal's functions would remain normal, Lashley wrote, "This may sound like a plunge into mysticism, but an example from another field will show that such self-regulation is a normal property of living things."[11] He went on to describe Child's experiments with crushing the tissues of hydroids, which showed that the formless mass of tissue underwent embryogenesis and ultimately produced normal structures. Later in the talk he referred again to Child's work, and remarked that the structure of the nervous system was such as to allow for the establishment of electrical or chemical gradients that would mediate brain function.

Lashley used the embryological analogy to argue against localization of function in the brain. Instead of specificity of location and isolation of structures, the embryological model emphasized patterns of excitation involving the entire cortex. Lashley told his audience, "We must look [to experimental embryology] for the next significant development in our knowledge of the functions of the brain." And to those who would criticize his analogy for its vagueness, Lashley responded, "We seem to have no choice but to be vague or to be wrong, and I believe that a confession of ignorance is more hopeful for progress than a false assumption of knowledge."[12]

A year later, in another address which subsequently appeared as an article in *Science,* Lashley again noted the usefulness of Child's work in explaining the phenomena of mass action.[13] Complexity of function was limited not by destruction of particular portions of the brain but by the sheer amount of functional tissue available. Citing Child, Lashley wrote, "A possible clue to the situation here comes from experimental biology. In the regeneration of hydroids the number of tentacles regenerated is correlated with the size of the regenerating mass of tissue."[14]

Lashley and Biological Determinism

Lashley's use of Child's gradient theory in particular and of embryological work in general to explain the complexities of brain function is proof of his debt to the

[10] Ibid., 171.

[11] Lashley, "Basic Neural Mechanisms in Behavior," *Neuropsychology of Lashley: Collected Papers of K. S. Lashley,* eds. F. A. Beach, D. O. Hebb, C. T. Morgan and H. W. Nissen (New York: McGraw Hill, 1960), 205.

[12] Ibid., 207.

[13] Lashley, "Mass Action in Cerebral Function," *Science* 73 (1931): 245–254.

[14] Ibid., 254.

American school. His principal divergence from that school's approach to psycho-biology is most evident in the position of the laboratory in his work. During the 1920s, at a relatively early phase of his career, Lashley asserted the laboratory as the preeminent place for the production of psychological knowledge.

In 1926, Lashley established his laboratory under the auspices of the Behavior Research Fund of Chicago's Institute for Juvenile Research. The institute was central to Progressive reform in Chicago, dedicated as it was to the more humane treatment of juvenile delinquents; it was associated first with the Cook County Juvenile Court and then with the Department of Public Welfare of Illinois. The Friends of the Institute, a group of Chicago citizens, established the fund in order to support basic research in behavior, and Lashley was one of their first appoint-ments. He gathered a productive group of students around him, both from the University of Chicago and from his department in Minnesota, and sponsored a series of researches which he subsequently collected in a volume called *Studies in the Dynamics of Behavior* (1932).[15] Lashley's introduction to this volume is especially revealing of the approach of his school.

At his first opportunity, in the first paragraph of the introduction, Lashley turned the Progressive agenda on its head. While the Progressives advocated reform, Lashley promoted its opposite: conformity. When Herrick spoke at the juvenile court celebration, he proclaimed that a better society could be evolved through individual will and collective action. Lashley, on the other hand, argued that individuals must be disciplined to uphold the "accepted social code." Rather than granting the individual the power to change himself and the system, Lashley looked for ways to make the individual fit within the system. "As yet," he wrote, "there are scarcely any recognized methods for developing the social attitudes which contribute to make the individual a useful member of society."[16]

Lashley's next point distinguished his approach even further from the Progres-sive program. He considered different individual temperaments – "the ambitious and the slothful, the dreamer and the go-getter, the honest and the dishonest, the selfish and the generous, the leader and the follower" – and identified the possible source of their differences: heredity or environment. He did not come down on one side or the other, but his statement of these alternatives clearly sets him apart from the American school.

The only way the relative importance of these two factors could be determined, Lashley concluded, was through laboratory experimentation. He acknowledged that laboratory studies were oversimplifications of social situations, but he re-garded them as necessary and – for the time being – adequate substitutes. Field studies in social control were not sufficient to solve the mysteries of behavior; the

---

[15] Karl S. Lashley, ed., *Studies in the Dynamics of Behavior,* by Calvin P. Stone, Chester W. Darrow, Carney Landis, and Lena L. Heath (Chicago: University of Chicago Press, 1932).

[16] Introduction, *Dynamics of Behavior,* vii.

scientific psychologist needed a laboratory where the conditions could be rigorously monitored and the variables isolated and investigated separately.[17]

Ultimately much study would have to be made of the ways in which the laboratory results could be applied in society. But first society had to be brought into the laboratory, inasmuch as that was possible, before the psychologist could presume to say anything about how temperament actually functioned in the world. That such laboratory studies were deemed relevant at all was a distinctive mark of the Lashleyan school. Lashley himself recognized the novelty of his assigning the laboratory a preeminent position with regard to society: the studies, he emphasized, comprised an "adventure in methodology." Pioneering a new method was more important, he said, than the "accumulation of masses of indecisive data."[18]

This new method – the laboratory investigation of temperament – had an inescapable effect on the science that got done. With social and environmental factors either dismissed or tightly controlled, the biology of temperament became the central topic of investigation. Lashley himself maintained that the social value and stability of temperament – its role in society – had to be approached through an understanding of its physiological correlates and genetic basis. This is hardly a self-evident claim: there is no clear or obvious reason that the physiology of emotion should shed any light on a delinquent's behavior in society. But for Lashley and many of his students, the link went unquestioned. They took the "biological point of view" because it could be most readily accommodated in a laboratory. With society eclipsed, the biological commonalities among all species loomed large. A study by Calvin Stone, for example, examined the genetic origin of temperamental differences and their modification by training; though Stone's subjects were rats, Lashley remarked that their "different heritable traits . . . [were] clearly suggestive of temperamental differences in man."[19] Human and animal studies were freely intermingled: in the following study, Chester Darrow and Lena Heath demonstrated correlations between physiological reactions and certain traits of personality. Only Carney Landis, examining the effectiveness of

[17] The approach of Lashley's students in *Studies in the Dynamics of Behavior* contrasts with the aims of contemporary social psychology, which was being organized as a field during the 1920s and 1930s. While the Lashley approach was to study the individual as much as possible in isolation from his society, the practitioners of social psychology always defined their field as the study of the individual in relation to his society. Like Lashley and his students, the social psychologists had an interest in quantitative and experimental methods, in rating schemes, questionnaires, and tests; but unlike Lashley, these students of social behavior focused on the interactions of human beings with each other, and on the collection, processing, and interpretation of mass data. See, for example, Floyd Allport, *Social Psychology* (Boston: Houghton Mifflin, 1924); Carl Murchison, *Social Psychology: The Psychology of Political Domination* (Worcester: Clark University Press, 1929); Gardner Murphy and Lois Barclay Murphy, *Experimental Social Psychology* (New York: Harper Brothers, 1931); and Richard T. LaPiere and Paul R. Farnsworth, *Social Psychology* (New York: McGraw Hill, 1936).

[18] Introduction, *Dynamics of Behavior,* viii–ix.

[19] Ibid., ix.

psychological tests as measures of emotional response, suggested briefly that perhaps physiology was not the best indicator of emotion and that sociological studies might be more appropriate. In Lashley's school, however, this suggestion was never followed up.

If Landis showed some deviation from his mentor's approach in his 1927 study, his dissertation work, completed three years earlier, is thoroughly imbued with the Lashleyan preeminence of the laboratory. In the dissertation, directed by Lashley at the University of Minnesota, Landis reconstituted society within the laboratory.[20] His method consisted of presenting human subjects with two dozen different situations that were supposed to elicit emotional responses. The subjects were asked to do everything from listening to music and reading Biblical passages to sniffing ammonia, decapitating live rats, and undergoing electric shock. The purpose, Landis said, was to create as many different types of reaction as possible, "working up in a cumulative manner so as to give a pronounced emotional upset."[21] Emotional responses among subjects could then be compared, facial expressions associated with certain emotions investigated, sex differences in emotional expression assessed, and so forth.

At no point in the study did Landis ever question the reality of the emotions he was provoking. Clearly the situations were artificial – rigged by the experimenter, stripped of any social context in order to be brought into the laboratory. But the emotions they produced were without doubt authentic; many of the subjects did indeed become clearly upset when confronted with certain of the situations; some even broke down in sobs and begged the experimenter to stop. "It seemed desirable," Landis wrote, "to arrive at some method by which real emotional disturbances could be engendered."[22] The fact that Landis believed he was creating such "real" emotions shows how completely the laboratory served as a substitute for society. The complicating social factors were simply dismissed, and the "real" response, the pure physical or physiological expression of emotions, reproduced within the laboratory's controlled environment. Human emotion did not require a society for its expression: it could be studied in exactly the same way as the behavior of a laboratory animal.

Landis brought society into the laboratory, and therefore under his control, in a material sense as well. A previous study of the emotions had revealed that subjects reacted as much to the laboratory, the apparatus, and the experimenter, as they did to the situation.[23] In order to lessen this effect, Landis tried to bring suggestions of

---

[20] Landis, "Studies of Emotional Reactions II: General Behavior and Facial Expression," *Journal of Comparative Psychology* 4, nos. 5 & 6 (October–December 1924). "Studies of Emotional Reactions I: A Preliminary Study of Facial Expression" appeared in the *Journal of Experimental Psychology* 7 (October 1924).

[21] Landis, "Emotional Reactions II," 453.

[22] Ibid.

[23] Landis, "Emotional Reactions I."

the laboratory to a minimum: he put paintings on the walls and curtains on the windows, and managed to conceal most of the apparatus. In this way society was simulated in the laboratory, both its material trappings and the reproduction of "real" emotions. Yet the experimenter's control was never forfeited, as it would have to be in the investigation of emotions in their social context.

We have so far marked two of the important moves that Lashley and his students made in psychology. First, they moved all investigation into the laboratory, thereby creating a standardized environment in which differences in their subjects' behavior could be observed. Second, through their control of environment they minimized "social" factors and focused entirely on the "natural," biological substrate of behavior. Their third move was an increasing identification of the biological with the innate.

This third aspect of the Lashleyan approach is clearly evidenced in the work of another of Lashley's students at Chicago, Frank A. Beach. Beach's dissertation, which he completed in 1937, was called "The Neural Basis of Innate Behavior" and featured the use of Lashley's technique of cortical destruction in rats to investigate the relationship of brain and behavior.[24] Beach identified maternal behavior, including giving birth and caring for young, as an innate pattern, and then destroyed different amounts and areas of cortical tissue to test the effects on this behavioral pattern. He observed that Lashley's principles of brain function obtained: the behavior deteriorated in proportion to the amount of tissue destroyed, and did not depend upon the preservation of any specific structure. The study also showed that innate behavior was just as much a product of the cortex as was learned behavior; that it was not controlled by subcortical structures, as influential contemporary belief held. For Beach, and for Lashley, this indicated more than a passing resemblance between innate and learned behavior; instinct and intelligence were deeply related through their dependence on the same neural structures. Instinct was therefore just as complex and as worthy of study as any learned behavior pattern. As Lashley's career progressed, the hereditary component of behavior began to loom ever larger in his work – the direct consequence of the preeminence of the laboratory, the marginalization of social and environmental factors, and the emphasis on the biological basis of psychology.

Herrick's Criticism of Lashley

Herrick objected to Lashley's approach every step of the way, not only to the hereditarianism at its endpoint, but also to the reductionistic biological determi-

---

[24] Frank Ambrose Beach, Jr., "The Neural Basis of Innate Behavior. I. Effects of Cortical Lesions Upon the Maternal Behavior Pattern in the Rat," *The Journal of Comparative Psychology* 24, no. 3 (December 1937): 393–439.

nism that supported it.[25] In his 1926 book *Brains of Rats and Men,* Herrick explored the connection between progressive evolution and creative intelligence, and ended with a criticism of Lashley's reductionism. Herrick noted his own reliance on Lashley's work, which had caused him to "recast" his "entire treatment of cortical function in mammals."[26] The first part of the book was devoted to the anatomy of the cerebral cortex and its evolution, the second part to mechanisms of learning in the rat, and the third part to the progress beyond the rat which the human species has made in consciousness and intelligence.

In the last section, Herrick's criticisms of Lashley became most pointed. Herrick opposed not only Watson's behaviorism, which denied that anything called "consciousness" actually existed, but also Lashley's reduction of consciousness to its physicochemical basis. For Herrick, Lashley's reductionism denied the power of mind to control events in the physical world. It denied the emergent psychological level at which mind appeared. Mind for Lashley became nothing but a special pattern of chemical and electrical transmission; its outcome in the world was no different from that of an instinct prompted by hormones. But instincts could not change the course of evolution; they were controlled by biology, not controllers themselves.

For Herrick mind was on a higher plane, no longer simply controlled or determined, but also in control. Therein lay his liberal hope for progressive change. About Lashley's reductionism, he wrote, "The demand that we evaluate scientifically subjective experience by exactly the same criteria as are adequate in physics and chemistry is a thoroughly unscientific procedure" because it denied the "personal quality" of mind in human experience, a quality that was missing in objectified natural processes. Herrick continued:

> No abyss of ignorance of what consciousness really is, no futilities of introspective analysis, no dialectic, destroy the simple datum that I have conscious experience and that this experience is a controlling factor in my behavior.[27]

No matter how complete our physiological explanations of psychological processes become, Herrick argued, they will never explain away our subjective awareness. Lashley's goal of reducing consciousness to physiology was therefore unattainable because "subjective experience remains a real fact of natural history."[28] To deny that thinking and choosing were processes that could control conduct was to deny Herrick's message of the possibility of social change. To

---

[25] On the Herrick-Lashley debate, see Sharon E. Kingsland, "A Humanistic Science: Charles Judson Herrick and the Struggle for Psychobiology at the University of Chicago," *Perspectives on Science* 1 (1993): 445–477.

[26] C. Judson Herrick, *Brains of Rats and Men,* ix.

[27] Ibid., 343.

[28] Ibid.

reduce these processes to the blind interactions of matter was to excise from behavior and evolution the possibility of conscious control. In Lashley's scheme, our conduct was determined by our biology, and we could not alter it because we could not stray from the path of biology. To this Herrick responded,

> It is a travesty of scientific method to leave out of consideration in a total view of human behavior just those characteristics which differentiate man from brutes and upon which the future progress of civilization must depend . . .

"Rats are not men," Herrick said in his concluding chapter; "men are bigger and better than rats."[29] Lashley's position was not only scientifically invalid (because it failed to take into account obvious facts of experience) but also morally dangerous. Reductionism, in making us creatures of our instincts and impulses, threatened to turn our hope for the future into an illusion by denying us the power to change our environment.

To drive this point home, Herrick never lost an opportunity to stress the differences in methods of learning between "higher" and "lower" animals. In lower animals, according to Herrick, learning was simply the transfer of habitual or instinctive behavior patterns to new situations: familiar responses to unfamiliar tasks. This type of learning typically required hundreds of repetitions before success was achieved. By contrast, in higher animals, Herrick continued, the learning process was completely different. A new behavior pattern or the use of an old one in a new situation could appear suddenly, in a flash of intuition. This type of behavior, Herrick wrote, "involves the use of a sort of neural organization quite different from that which is adequate for the simpler conditioned responses . . ."[30] Later, in discussing the localization of the learning process in the rat's cerebral cortex, Herrick wrote:

> One can agree with Lashley . . . , "The rat has a very primitive cerebral organization but I doubt that this justifies the assumption that there is any fundamental difference in cerebral mechanism between the rat and even man"; but only in the light of a remark appearing lower on the same page, "The differences between the rat and the apes are scarcely greater than those between the apes and man."

He continued, "It all depends upon what we mean by 'fundamental difference.' The difference is certainly significant . . ."[31]

Lashley and Herrick's disagreement over evolutionary progress was a symptom of a broader disagreement over the definition of science, or, more precisely, over

---

[29] Ibid., 345, 347, 365.
[30] Ibid., 65–6.
[31] Ibid., 214–15.

the definition of mechanism. Lashley and Herrick each believed that his own science was in accord with the strictest principles of mechanism; but they could not agree as to what those principles were. For Lashley, anything short of biological (and, ultimately, genetic) determinism was simply not science. Consciousness and intelligence could be explained only as products of the brain, the organization and activity of which were in turn determined by the genes. This was the most fundamental level in the reductionist hierarchy. Inserting some immaterial entity like "free will" destroyed the causal continuity from gene to behavior. Herrick, on the other hand, believed that this reductionism denied a fact of existence, the capacity of the mind to control behavior, and so ran contrary to the principles of mechanism. He wrote,

> I claim to be a more radical mechanist than those apostles of behavior-ism and objective psychology who spurn that introspective experience through which alone they are able to codify and interpret their objective experience.[32]

Both Herrick and Lashley, then, were biological determinists, but in radically different modes. Their debate turned on two major issues: first, on progress in nature; and second, on the extent to which natural laws can provide prescriptions for society. For the Progressive psychobiologist, signs of progress were every-where in the natural world – in embryology, in evolution, in the mental develop-ment from rats to men. From such evidence of biological progress, Herrick made a seamless transition to his hopeful ideal of social betterment: the laws of nature clearly showed that society too was on an upward path toward greater democracy.

But for Lashley, Herrick's vision of progress in nature was an illusion. The anti-Progressive argued instead for the constancy of neural mechanisms throughout the mammalian species, thereby denying that progress either had occurred in the past or should be expected in the future. And he opposed the notion of an emergent level of mind: there was no way for a rat or a man to rise above the level of biology. Biology could not be changed by "will power."

Moreover, Lashley broke with the Progressive psychobiologist's analogies be-tween science and society. The research findings of neuropsychology had, Lashley contended, no bearing on social problems. Though I will argue in a later chapter that Lashley, in his own way, did in fact take prescriptions for society from his scientific work, all along he denied the social relevance of his experi-ments. This belief in the social meaning of science, which Herrick held so dear and Lashley so firmly denied, was a major point of contention in the debate between the Progressive and the anti-Progressive.

---

[32] Herrick, *George Ellett Coghill, Naturalist and Philosopher* (Chicago: University of Chicago Press, 1949), 222.

Tensions at the Behavior Research Fund

But there were more than two sides to this debate over the political implications of psychology and the relationship of psychology to society. The easy transition that Herrick made from biological problems to social issues was a problem not only for Lashley; it was also source of much dissent among the social scientists of the Behavior Research Fund. The psychiatrist Herman Adler, the director of the fund in its early years, expressed well the tensions between those who, like Herrick, saw a biological foundation to social problems and those who believed in cultural explanations for such problems. Lashley's monograph was the first to be published by the University of Chicago Press under the auspices of the fund. In his foreword Adler wrote:

> It may puzzle some that the approach to human behavior should lead through such an apparent bypath, but a frontal attack is not necessarily the most effective. Nature's secrets sometimes yield to an ambuscade.[33]

Yet that same year, 1929, Adler also wrote the foreword to the second of the fund's publications, *Delinquency Areas: A Study of the Geographic Distribution of School Truants, Juvenile Delinquents and Adult Offenders in Chicago,* by Clifford Shaw. Here Adler said that

> when the results of animal experimentation are presented, the critic discovers that the application of the results to human beings is not valid. This volume deals with experiments which were performed by the circumstances of life itself, and not by an experimenter.[34]

A research sociologist at the institute, Shaw himself wrote that his approach to juvenile delinquency was a cultural one; while he realized that there might be a physiological component to the delinquent's behavior, his study was not concerned with it. Rather, Shaw set out to understand the cultural and social factors that contributed to delinquency: the individual's family life, his perception of the world, the parts of the city and the groups of people toward whom he gravitated, his early experiences with crime and sex. Adler noted that such a cultural approach was leading to the "establishment of a social science" that was quite separate from biological or genetic factors. Shaw made no attempt to trace any sort of delinquent factor through families, nor any suggestion that the offenders were biologically predisposed to commit crimes. He stressed therapy for these criminals and suggested that they be permitted to marry, have children, and carry on normal lives.

In fact, an approach that Shaw developed from the work of the sociologist William I. Thomas and made famous in his book *The Jack Roller* was that of the

---

[33] Herman Adler, foreword, *Brain Mechanisms and Intelligence,* by Lashley, viii.

[34] Clifford Shaw, *Delinquency Areas* (Chicago: University Of Chicago Press, 1929), vii.

delinquent's "own story."[35] The delinquent was supposed to write the history of his own career; these accounts were then published, with few changes, in the delinquent's own language. *The Jack Roller,* which is the story of Stanley, is perhaps the most famous example of this genre; at the end Stanley has not only changed his evil ways but has a wife and child.

These "own story" accounts served a dual purpose. First, they allowed the sociologist to get inside the delinquent's head, to see the world as he saw it, not because this would provide an accurate picture of the world, but because it would give greater insight into the workings of the delinquent's mind. The second purpose was to show that recovery was entirely possible: the delinquents had to regain their hold on normal life and achieve some detachment from their troubles and crimes in order to write about them. These stories showed that beneath the hard-drinking, sexually deviant criminal exterior there was a real human being with thoughts and feelings and the hope for a normal life. The sociologists who collected and edited these stories suggested that a normal life was well within the grasp of these offenders, and that their criminal behavior was a product of social and cultural circumstance. In his commentary on the jack roller's story, in fact, Shaw concluded that a criminal's disposition was the result of his upbringing and his experience, not his biological or genetic makeup. His biology, Shaw believed, was more or less incidental; society was the real causal factor in his behavior:

> The human being as a member of a social group is a specimen of it, not primarily, if at all, because of his physique and temperament but by reason of his participation in its purposes and activities. . . . Therefore the relation of the person to his group is . . . on a cultural rather than upon a biological level.[36]

This was a strong environmentalist position. Shaw justified his approach to his subject and his own kind of science by giving a cultural genealogy to the delinquent's problems.

Adler's forewords to the first two Behavior Research Fund Publications, Lashley's and Shaw's, marked a tension in the research carried out under the auspices of the fund. The biological formulation of behavior problems by Lashley and his students was opposed to the sociological approach. This tension remained during the lifetime of the fund: while Lashley and his students carried out animal research, and other psychologists at the Fund cited their results, the sociologists pursued their own direction. Lashley's edited volume, *Studies in the Dynamics of Behavior,* which took "the biological approach," freely mixing animal and human results, appeared in 1932 and affirmed exactly what *The Jack Roller* had denied –

[35] Clifford Shaw, *The Jack Roller: A Delinquent Boy's Own Story* (Chicago: University of Chicago Press, 1930). This book was the fifth of the Behavior Research Fund Publications.
[36] Shaw, *The Jack Roller,* 186.

that temperamental differences were biologically determined and were crucial factors in determining behavior.

A resolution of this tension was not suggested until 1933, by Lashley's friend and associate Heinrich Klüver. In his book *Behavior Mechanisms in Monkeys,* Klüver essentially took Lashley's approach to its extreme: he advocated a complete separation of the biological and the cultural approaches. But instead of making either implicit or open comparisons between rats and men, as Lashley had done, Klüver argued that we must try to strip away all sociological meaning from our studies of animals to reveal the biological substrate.[37] His study, he believed, was thoroughly biological; no application to the problems of society was meant to be stated or implied. (Whether this is what he actually achieved is another matter.) In Klüver's book the biological/cultural tension results in a complete bifurcation of the two approaches. Klüver's work was done under Lashley's supervision; but two years later Lashley moved to Harvard, and the Behavior Research Fund published nothing else relating to animal research after Klüver's work on monkeys.

The tension between biological and cultural determinants of behavior appeared again in Lashley's debate with Clark Hull. While Herrick and Lashley agreed on the biological substrate of behavior, and made this agreement the basis for a collaboration, Hull and Lashley differed radically on this issue. Yet Hull and Lashley were both opposed to the compromised mechanistic psychology they believed Herrick's psychobiology represented. Lashley's dispute with Hull was finally even more unresolvable than his debate with Herrick, and it illustrates the opposing directions in which psychology was being pulled during the 1930s, 1940s, and 1950s.

---

[37] Heinrich Klüver, *Behavior Mechanisms in Monkeys* (Chicago: University of Chicago Press, 1933), xi–xiii.

7

Hull and Psychology as a Social Science

Prologue

In *The Triumph of Evolution,* the historian Hamilton Cravens argued that by 1941 many natural and social scientists working in the United States had managed to resolve their ancient dispute over the relative values of nature and nurture in development. Heredity and environment were interdependent, cooperating factors, they concluded; their effects were not separable, nor was one more important than the other.

Earlier in the century, according to Cravens, most biologists, psychologists and sociologists had tended to come down on one side or the other, stressing the role either of innate constitution or of learned behavior. But by 1941 they had "recognized the complexity of human nature and behavior" and declared the nature-nurture dichotomy "artificial, unproductive and perhaps unscientific."[1] Eager to take up new questions, life scientists formulated a new interactionist model: on the basis of this new synthesis they forged a science of man, a truly interdisciplinary study of humanity that took into account both biological and cultural evolution. This triumph of evolution and resolution of the controversy ushered in an era of interdisciplinary scholarship.

The resolution of this longstanding controversy, however, was far from a universal conclusion. While many prominent life scientists did indeed pronounce a harmonious end to the dispute, in less visible or less public arenas the old debate continued to rage unabated, and continued to exert a powerful influence on the sciences of life and mind.

For example, in 1947, six years after natural and social scientists had supposedly brought the nature-nurture debate to closure, *The Psychological Review* devoted an issue to the "Symposium on Heredity and Environment" presented at the annual meeting of the Society of Experimental Psychologists at Princeton in April of that year. At this symposium, papers were presented by Karl Lashley, by Lashley's students Frank A. Beach and Calvin P. Stone, and by Lashley's associates Leonard Carmichael and Clifford T. Morgan. In his summary comments on

[1] Hamilton Cravens, *The Triumph of Evolution: American Scientists and the Heredity-Environment Controversy* (Philadelphia: University of Pennsylvania Press, 1978), 251, 264.

119

the symposium, the psychologist Walter S. Hunter observed, "It will come as a surprise to many psychologists that all five of the distinguished contributors to this Symposium have emphasized the role of heredity in the determination of behavior." This was a surprise, Hunter explained, not because the debate had already been resolved, nor because the scientific community was moving on to other issues, but because during the previous twenty-five years so many psychologists had been staunch environmentalists. Hunter himself defended an environmentalist position, arguing that except for genetic control of reflexes, "evolutionary development has largely freed man from a dependence upon directly inherited behavior patterns."[2]

The example of the Heredity-Environment Symposium indicates that beneath the surface of conflict resolution and new beginnings so confidently announced by certain leaders in American science, there was still considerable discord and disagreement about how behavior was determined, whether by inborn propensities or by environment. For Lashley and the community of scientists he led at the symposium, the nature-nurture dichotomy remained a live issue; so much so, in fact, that it became the source of Lashley's fifteen-year, unresolved and unresolvable controversy with the neo-behaviorist Clark Hull.

A central event of Lashley's middle years, in the 1930s and 1940s, was his debate with Hull over the structure and function of the brain, its relationship to the mind, and the use of machine metaphors to explain intelligence. On the surface, their dispute was not about the relative importance of heredity and environment, but that subtext ran through all their exchanges. Far from agreeing that the nature-nurture controversy had been resolved and deciding to move on to other problems, Lashley and Hull committed themselves to opposite sides of the issue. Their positions in turn influenced their outlooks on a host of other problems, even those that might appear at first to have little connection with that debate. Far from agreeing that a synthesis of heredity and environment should form the basis for their science, Lashley and Hull differed precisely on the proper foundation for a science of man.

A determined hereditarian, Lashley was committed to the notion that innate intelligence could be measured and to a belief in the biological foundation of mind. These commitments led him to espouse a dynamic view of brain function, a view which brought him into direct conflict with the neo-behaviorist Hull. Hull brought to its fullest development Watson's conception of mechanical man, believing both that the workings of the brain could be mimicked by a machine, and that machines were capable of displaying genuine intelligence. His theory that environmentally shaped reflex connections underlay all behavior was the centerpiece of Yale University's Institute of Human Relations, its program of unifying

---

[2] W. S. Hunter, "Summary Comments on the Heredity-Environment Symposium," *Psychological Review* 54 (1947): 348, 351.

the social sciences and its aim of developing more efficient methods of social control. Lashley argued against Hull's use of the machine metaphor for the mind and against his notion that stimulus-response connections could be traced through the brain and nervous system. He also objected, on a more fundamental level, to Hull's unification of the social and behavioral sciences under his own banner, and to the environmentalist assumptions that underlay his whole system.

This controversy demonstrates, however, that the heredity/environment split was not a simple dichotomy. Lashley did not believe that every mental and physical trait was completely genetically determined, nor did Hull think that behavior was infinitely modifiable by environment. I call Lashley and Hull respectively hereditarian and environmentalist because that is the least prolix characterization of their standpoints. But both Lashley and Hull argued for positions that were actually compromises between the two extremes: there was room for some environmental variability in Lashley's theory, and Hull conceded that innate differences did play a role in shaping behavior.

Lashley and Hull, then, each worked out his own resolution to the heredity/environment debate; yet they were still unable to agree. Neither was an extremist, yet they chose moderate positions that were nevertheless radically different. Their dispute demonstrates that scientists need not be extremists to disagree vehemently, and that there is no one logical and foreordained conclusion to a controversy. The Lashley-Hull debate remained unresolved because each participant held a position that made perfect sense to him, to which he felt driven by his experimental evidence, and which fit with his broader intellectual and institutional commitments.

The controversy also shows that these psychologists' theories of how the brain works to produce the mind were inseparable from their ideas about intelligence. Lashley's hereditarianism – his belief that each person was endowed with a certain degree of intelligence which she could do little, if anything, to alter – produced a completely different theory of brain function from the one that Hull endorsed from his environmentalist standpoint.

This chapter shows how Hull continued the behaviorist tradition of distinguishing psychology from biology and of stressing the role of the environment in shaping behavior. By proposing a basic law underlying behavior, Hull's neobehaviorism made psychology the unifier of the social sciences. In the next chapter, I will examine the actual interactions between Lashley and Hull.

Hull's Background

Clark Hull (1884–1952) was trained in psychology at the Universities of Michigan and Wisconsin; after earning his Ph.D. in psychology at Wisconsin he remained there as a member of the department of psychology. His interest in the conflict between behaviorism and Gestalt theory was sparked by a visit to Wisconsin of the German psychologist Kurt Koffka; but instead of becoming sympathetic to the Gestalt standpoint, Hull was inspired to find a way to shore up the behaviorist system. This took him the rest of his career. In 1929 he moved to Yale University as research professor in the Institute of Psychology, which later that year became incorporated into the Institute of Human Relations.[3]

The differences between Lashley's training and Hull's are apparent. Lashley was never officially trained in psychology and came to it only later by way of his interest in biology. Hull was drawn to psychology because it allowed the possibility of using mechanical devices in building its theories. He had originally wanted to become an engineer, but decided after a crippling polio attack that he was too "feeble" for it, and so chose psychology as the next closest profession.[4] He did in fact invent a number of machines, among them an automatic memory machine, a machine that could calculate aptitude measurements, and a machine that mimicked a conditioned reflex. Hull's interest in machines also profoundly separated him from Lashley.

Hull chose psychology also because, as he said, it offered opportunities for the theoretically inclined, again in contrast to Lashley, who proclaimed a great disregard for theory. Hull contended that in such a new field as psychology, a young investigator could make a contribution to theory and a name for himself without having to wait for older, well-established predecessors to die.[5] An indication that Hull's consuming passion was theory is Frank Beach's comment about him that he "never conducted experiments himself."[6] This statement has been disputed by other commentators on Hull's life, but regardless of its accuracy, it certainly could never have been applied to Lashley. Finally, unlike Lashley, Hull spent his career in one place after his stint at Wisconsin: Yale's Institute of Human Relations.

Hull's Neo-Behaviorism

Hull was a leading neo-behaviorist theoretician. Following Watson, Thorndike, and especially Pavlov, he believed that all behavior was built up from reflex connections between stimulus and response, and that these reflexes could be conditioned – that is, modified or altered – by changing the environment. Unlike

---

[3] The details of Hull's life can be gleaned from his autobiography, "Clark L. Hull," *A History of Psychology in Autobiography*, vol. 4, eds. E. G. Boring, H. S. Langfeld, Heinz Werner, and R. M. Yerkes (Worcester: Clark University Press, 1952), 143–62; from Frank A. Beach, "Clark Leonard Hull," *Biographical Memoirs of the National Academy of Sciences* 33 (1959): 125–141; and from Laurence D. Smith, *Behaviorism and Logical Positivism: A Reassessment of the Alliance* (Stanford University Press, 1986), 147–184.

[4] Smith, 150.

[5] Ibid.

[6] Beach, "Clark Leonard Hull," 133.

the others, however, Hull was more self-consciously creating a theoretical system: he argued that in the conditioning of reflexes he had found the fundamental law of behavior, and that this law could form the basis for a unified social science. He expressed his belief in this fundamental law at the end of his autobiographical essay:

> Perhaps the most effective means to that great end [ridding both science and society of subjectivity] will be the accurate and wholly convincing deter- mination of the primary laws of human behavior. . . . These laws should take the form of quantitative equations readily yielding unambiguous deductions of major behavioral phenomena, both individual and social.[7]

Hull's basic law of behavior, which was supposed to form the foundation for all the social sciences, was the law of stimulus generalization. Simply put, this law stated that a response could be called forth by an unconventional stimulus as long as that stimulus was associated, either temporally or in character, with the stim- ulus that usually called forth the response. The central feature of the theory was that responses could be called forth by engineering the appearance of stimuli. As long as the unconventional stimulus was similar enough to the usual one, it could call forth the response. So, for example, any other stimulus that happened to occur at the same time as the main stimulus would also in subsequent trials provoke the same response; Pavlov had noticed this phenomenon when his dogs salivated at the ringing of a bell.

From this mechanistic, automatic process, Hull could derive complex and adaptive behavior without appealing to consciousness or choice on the part of the organism. The law of stimulus generalization explained how all learning took place, without resorting to immaterial forces such as a soul or free will. No spirit in the organism decided what responses should answer a stimulus; in fact, nothing at all internal to the organism determined how it would learn. The learning process was for Hull wholly regulated by the environment; it was the source of the modifications in behavior that constituted learning. The organism merely had to respond in a stereotyped way.

Hull's discovery of this basic law of behavior was the centerpiece of the Institute of Human Relations – not only its crowning achievement, but the piece of work that justified the institute's existence and determined the character and direction of the rest of the research carried out under its auspices. The life of the institute was coextensive with Hull's tenure at Yale: it opened in 1929, the year he was appointed, and it was phased out when he died in the early 1950s. During that period he was its guiding spirit. It is difficult to imagine an individual and an institution better fitted to each other. In his 1943 book *Principles of Behavior*, Hull

---

[7] Hull, "Clark L. Hull," 162.

wrote that his was "the theory of all the behavioral (social) sciences"; the director of the Institute, Mark May, wrote seven years later:

> The first step toward the achievement of a unified inter-disciplinary science of behavior and human relations is the discovery of one or more integrating principles which appear, tentatively at least, to have been valid for all men all the world over at all times since the mammalian species appeared.[8]

With his principle of stimulus generalization, Hull gave the institute such a universally applicable law.

Hull's theory rested on his redefinition of behavior. In order to understand the mechanism of learned reflex response as the regulator of the rat's behavior in the maze, of the juvenile's penchant for delinquency, of the activities of the Siriono people of lowland Bolivia, Hull had to redefine behavior as the outward response displayed by an organism to a stimulus, and to exclude rigorously from his formulations all reference to the biological basis. While Jennings tried to include in "behavior" both overt response and physiological activity, Hull defined two distinct categories of behavior, molecular and molar, and focused his study on the latter. In a footnote to his 1943 *Principles of Behavior,* he wrote:

> By this expression [molar] is meant the uniformities discoverable among the grossly observable phenomena of behavior as contrasted with the laws of behavior of the ultimate "molecules" upon which this behavior depends, such as the constituent cells of nerve, muscle, gland and so forth. The term *molar* thus means coarse or macroscopic as contrasted with *molecular,* or microscopic.[9]

Likewise, Mark May noted in his 1949 survey of the institute's work that

> a large body of scientific facts and theories from the fields of bio-chemistry, neurophysiology and neuroanatomy have not been included in the integration described in the first section of this chapter. There are two reasons for this. One is that not enough is known as yet about the biochemical nature of the innate mechanisms of learning to help the behavioral scientist solve his problems. The second reason is that by using the "postulational-deductive-experimental approach" [of Hull] the scientist can proceed with the development of basic principles without waiting for exact knowledge of their inner nature.[10]

---

[8] Mark A. May, *Toward a Science of Human Behavior: A Survey of the Work of the Institute of Human Relations Through Two Decades, 1929–1949* (New Haven: Yale University Press, 1950), 23.

[9] Hull, *Principles of Behavior: An Introduction to Behavior Theory* (New York: D. Appleton-Century, 1943), 17n.

[10] May, *Toward a Science,* 10. "Postulational-deductive-experimental" is synonymous with "hypothetical deductive."

The Social Sciences at Yale

Yale's Institute of Human Relations evolved from the Institute of Psychology, the plan for which was conceived in 1921 by James Rowland Angell, then the president of Yale. A psychologist trained in the school of Chicago functionalism, Angell envisioned the Institute of Psychology in the image of Chicago's department of psychology: that is, he believed that the study of biology was crucial to understanding human behavior. Under his direction, the institute became the home base for the National Research Council's Committee for Research on Problems of Sex, and for the Committee on Human Migration: both had mandates to study the biology as well as the social consequences of sex and race. The first appointments in the new Institute of Psychology included not only the experimental psychologist Raymond Dodge and the anthropologist Clark Wissler, but also Robert Yerkes to represent psychobiology.

The appointment of Yerkes, who was a strong supporter of the integration of psychology and biology, and his subsequent brief career at the institute are telling indications of the institute's original plan of research and the evolution of that plan. Yerkes remained at the institute only until the early 1930s, by which time it had become the Institute of Human Relations. By 1932, Yerkes had become a member of the Department of Physiology of the Yale School of Medicine and director of the Laboratories of Comparative Psychobiology.

The point is that as the Institute of Psychology became the Institute of Human Relations, it deviated (in practice, at least, if not in theory) from Angell's original intention by reducing the role of biology in its research plan. Despite the fact that the Rockefeller Foundation Annual Report for 1929 included Yerkes's laboratory as part of the Institute, Yerkes considered his primary affiliation to be with the department of physiology. He did not have an open disagreement with the director or members of the Institute of Human Relations; he and the institute were simply heading in different directions. The physical plant of the institute also reflected the place of biology in it: the new institute building housed the social science departments, psychiatry, psychology, and the Child Development Clinic; the affiliated departments of biology were housed in an adjoining building.[11]

The purpose of the Institute of Human Relations, which was officially founded in 1929 with support from the Rockefeller Foundation, was to "coordinate and invest with a certain physical unity" the various sciences dedicated to the study of man, both as an individual and in society. Psychology, psychiatry, sociology, and anthropology were to be unified on common principles. The institute drew its members from the former Institute of Psychology and its child study clinic; from workers in mental hygiene; from the psychiatrists at the School of Medicine; and

[11] Mark A. May, "A Retrospective View of the Institute of Human Relations at Yale," *Behavior Science Notes* 6 (1971): 146.

from social scientists in anthropology, sociology, economics, and the School of Law. "The normal individual is to be studied with the same care as has been commonly applied to the abnormal; group, as well as individual, behavior is to be recognized as significant," the Rockefeller Foundation report explained in 1929.

The institute's new building was constructed at Yale to provide laboratory and library space for its members (except for those in biology), thereby reinforcing the institute's role in integrating and unifying the various disciplines. The institute was directed at first by an executive committee composed of administrative officers of the university, including President Angell; but when Angell retired in 1937, the membership of the committee shifted to the research staff of the Institute, including Hull, with Mark May as director.[12]

May and the other leading members of the institute saw Hull's work as the heart of the institute's integration of the social sciences. Hull's theory of behavior linked psychology, psychiatry, sociology, and anthropology by describing the mechanism of learning as the forging of connections between stimulus and response, and then envisioning this mechanism as the mediator of all social and cultural activity. The anthropologist George Peter Murdock, for example, wrote that the various social sciences actually formed a single integrated science based on Hull's principle:

> Since society, culture and personality are all learned, students of any of these phenomena must constantly bear in mind the fundamental principles of learning as these have been worked out by such behavioristic psychologists as Hull. Unless they do so, their conclusions will suffer in clarity if not in validity.[13]

Mark May remembered Hull's Wednesday Evening Seminars on learning theory as a source of much of the institute's integrative work. During 1935 and 1936, for example, they were focused on trying to integrate learning theory with psychoanalytic theory. John Dollard, an anthropologist with training in psychoanalysis, collaborated with Hull and several other members of the institute on this project. "The leaders of the meetings were convinced," May wrote, "that psychoanalysis and stimulus-response psychology were not mutually incompatible."[14] May also credited Hull's Wednesday seminars with working toward the integration of the laws of behavior with the work of various social theorists, including Marx and Sumner. And in the realm of cultural anthropology, Hull's point of view was also well represented; May described how Miller and Dollard in their book *Social*

---

[12] On the institute, see J. G. Morawski, "Organizing Knowledge and Behavior at Yale's Institute of Human Relations," *Isis* 77 (1986): 219–242.

[13] Murdock, "The Science of Human Learning, Society, Culture and Personality," *Toward a Science of Human Behavior,* 74.

[14] May, "A Retrospective View of the Institute," 163.

Learning and Imitation (1941) "emphasized and illustrated how the principles of learning operating under the conditions of social life produce social habits."[15]

As director of the institute, May certainly had an interest in emphasizing how well the integration worked, and what a variety of studies were combined on common principles. What is notable about his retrospective view of the institute is the prominent role that Hull plays in his descriptions of its work: Hull was the unifying point, the person who brought together diverse theories, who provided in his seminars an opportunity for interdisciplinary discussion. It was also Hull, according to May, who suggested a way of holding things together when they seemed ready to fly apart. In the spring of 1935, May announced that the problem of finding a common ground for the institute's diverse researches had reached crisis proportions, and called for proposals to coordinate the work. Hull's proposal, May noted, was the one taken most seriously: he suggested that the institute's work be centered on the problem of "motivation," and that this be the unifying theme. The fate of Hull's proposal is telling:

> This proposal was seriously considered for several months, until it became apparent that a majority of the members of the group were really not interested in it. In spite of a strong desire for the prosperity and continuation of the Institute . . . it seemed to most of [the members] . . . to be wasteful, if not utterly futile, to attempt to reorient their work toward the problem of motivation.[16]

The institute was ultimately dissolved in 1952; Mark May reported that its interdisciplinary structure had fallen out of favor with Yale's administration.[17] Whether by that point its work had been completed, and whether that work was a success or a failure, are not my concerns here. For my purposes, what the institute's story does show, strikingly, is how utterly opposed Hull's outlook on psychology was to Lashley's. For Hull and his institute colleagues, psychology was a social science, focused on the outward behavior of the individual and his relations to his environment. Hull's perspective could not have been more different from Lashley's biological, brain-centered neuropsychology. It should be no surprise, then, that the clash that resulted from these sharply differing points of view occupied both Hull and Lashley to the ends of their lives.

[15] Ibid., 165–6.
[16] Ibid., 157.
[17] Ibid., 168.

8

Intelligence Testing and Thinking Machines: The Lashley-Hull Debate

O Nature, and O soul of man! how far beyond all utterance are your linked analogies! not the smallest atom stirs or lives in matter, but has its cunning duplicate in mind.

Melville, *Moby-Dick*

Intelligence Testing and the Machine Analogy

Hull's conflict with Lashley took place on three levels. On the most obvious level, theirs was a dispute about the structure and function of the nervous system. Lashley believed the brain functioned as an equipotential system, as a whole, and that a certain amount of brain tissue was necessary, regardless of its location, for proper functioning. For Hull, the nervous system was a mosaic of discrete connections between stimulus and response, each particular connection activated by an impinging stimulus.

On a disciplinary level, they clashed about the proper relationship of psychology to the other sciences: for Lashley, psychology was completely reducible to the biology of the brain and closely connected to neurology and embryology. But Hull believed that psychology was the most basic of the social sciences and should work toward their unification, without reference to the "molecular" sciences. And on the most basic level, they were divided on the issue of the relative importance of heredity and environment in determining behavior. For Lashley, the locus of control of behavior was central: the innate constitution of the brain determined behavior. For Hull, the locus of control was peripheral: the organism's behavior could be modified by changing its environment.

Their basic dispute over heredity and environment is illuminated by their different approaches to intelligence testing. Echoing and reinforcing Spearman's doctrine, Lashley believed that intelligence was a unitary factor corresponding to the amount of functional brain tissue. For Hull, however, there was no such thing as a general intelligence which had a direct neurological correlate. In Hull's conception, intelligence was nothing but the average of discrete and specific aptitudes, each of which was as much environmentally determined as it was hereditary. Because the notion of a general intelligence was the banner of the hereditarians, Hull disavowed it completely; he was even unwilling to call what his tests measured "intelligence," preferring instead to say diverse "aptitudes." "Among testing experts," Hull wrote, "there is a growing tendency to admit that what have usually been called 'general intelligence' tests are in reality tests of scholastic aptitude, i.e. a kind of general average of the various aptitudes for learning the different school subjects." He dismissed the notion of intelligence as unscientific: "The recognition that if a test is to be of any particular value it must enable us to forecast a *particular* aptitude or group of aptitudes rather than measure some hypothetical or semi-metaphysical faculty, constitutes a great advance."[1]

Hull objected strenuously to the notion that individuals could be ranked in unilinear succession according to the amount of "intelligence" they possessed, or that they could be classified into two groups, the "intelligent" and the "feeble-minded." Differences in capacity within a single individual were often greater than differences between individuals, he argued. Like Spearman, Hull's purpose in using intelligence tests was to guide individuals to professions for which their capacities were best suited; but unlike Spearman, he was not looking to neuropsychology for any corroboration of his concept of aptitude. Spearman believed that Lashley's discoveries had proved the existence of a general intelligence. But Hull disregarded biology in his determination of aptitude. All his evidence came from his tests: he believed that nothing could be known of the essential nature of an individual's aptitudes. Simply by using the results of the tests "it clearly would be possible to tell within limits whether a given subject would have greater ability in aptitude A or aptitude B merely from knowing the scores in tests a and b. In other words, the existence of group factors would permit the possibility of differentiating the potential aptitudes of an individual by means of tests."[2] Hull believed that correlations between aptitude and physiological data were neither necessary nor possible.

Hull's use of the machine analogy also helped him to drive a wedge between psychology and biology. He believed that the human body operated like a mechanical apparatus, and that the day was not far off when machines would be built that would display genuinely intelligent behavior. The nervous system, after all, was like a telephone switchboard, automatically making connections between incoming and outgoing lines. For Hull, then, there was nothing special about the biology of organisms that made them capable of intelligent action: such action could be precisely mimicked by nonorganic mechanisms. Nothing inherent in biological organization produced this capacity; and so the study of biology would not lead to a better understanding of intelligence.

Both Hull and Lashley were deeply anti-idealistic in their approach to psychol-

---

[1] Hull, *Aptitude Testing* (N.Y.: World Book Co., 1928), 19.
[2] Ibid., 204–5.

ogy; their materialism led them to argue against anything that could be the last outpost of the soul. But for Hull, the notion that intelligence was solely a property of biological organization was the equivalent of granting it mysterious metaphysical characteristics: it was a back door through which the soul could creep. Ever vigilant against this possibility, Hull designed a machine that showed the phenomenon of rote learning; another that worked on the principle of the conditioned reflex; and, as Laurence D. Smith has pointed out, Hull viewed his whole theoretical system as a kind of machine from which intelligence could be deduced.[3] Whether the machine parts were mechanical or theoretical did not matter: they were equally efficacious in "dissolv[ing] the age-old problem of the opposition of mind to matter."[4] Hull remarked:

> [I]t should be a matter of no great difficulty to construct parallel inanimate mechanisms, even from inorganic materials, which will genuinely manifest the qualities of intelligence, insight and purpose, and which, in so far, will be truly psychic. . . . That such mechanisms have not been constructed before is doubtless due to the paralyzing influence of metaphysical idealism. The appearance of such "psychic" mechanisms in the not very remote future may be anticipated with considerable confidence.[5]

In fact, Hull contended that the construction of "psychic machines" would provide a "shortcut" to understanding the materialist basis of mind without taking a long detour through biology. Describing the laws by which, he argued, behavior is learned, Hull wrote:

> Suppose it were possible to construct from inorganic materials . . . a mechanism which would display exactly the principles of behavior presented in the six postulates just examined. On the assumption that the logic of the above deductions is sound, it follows inevitably that such a "psychic" machine, if subjected to appropriate environmental influences, must manifest the complex adaptive phenomena presented by the theorems. And if, upon trial, this a priori expectation should be verified by the machine's behavior, it would be possible to say with assurance and a clear conscience that such adaptive behavior may be "reached" by purely physical means.[6]

In the late 1920s and 1930s, Hull worked closely with chemists to design machines that could carry out the functions of a living organism. In 1929, he and H. D. Baernstein, a physiological chemist at the University of Wisconsin, announced their construction of a mechanical parallel to the conditioned reflex.

---

[3] Smith, *Behaviorism and Logical Positivism*, 168.

[4] Hull, "Simple Trial and Error Learning: A Study in Psychological Theory," *Psychological Review* 37 (1930): 255–6.

[5] Ibid.

[6] Hull, "Mind, Mechanism and Adaptive Behavior," *Psychological Review* 44 (1937): 29n.

"Learning and thought are here conceived as by no means necessarily a function of living protoplasm any more than is aerial locomotion," they wrote.[7] In 1931, a more detailed description of the apparatus appeared in the *Journal of General Psychology*.[8] The same year Hull collaborated with Robert J. Krueger on an electrochemical parallel to the conditioned reflex.[9] Hull saw this work as part of a trend toward the design of intelligent machines. Indeed, later in the decade, Douglas G. Ellson built a mechanical model of trial and error learning; and the Chicago Century of Progress Exhibition featured "mechanical men" which explained to visitors the physiology and chemistry of the human body.[10]

Hull's view of the human organism was, then, shaped by the machine analogy, and it led him to think that both the individual and society could be engineered, just as actual machines were. Applied psychologists were social engineers, he declared, and could work from his own pure psychological principles just as actual engineers worked from the principles of physics. The achievement of the engineering potential implicit in Hull's theory depended on a close relationship between pure psychologists (Hull and his school) and clinical and aptitude psychologists, penologists, psychiatrists, sociologists, and cultural anthropologists. "This would be in accordance with the practice of engineers who now receive training in the basic disciplines of mathematics, physics and chemistry," Hull wrote.[11]

In contrast to Hull, Lashley was deeply opposed to the machine analogy. By offering an equally materialistic alternative to the study of biology and neurology, the intelligent machine rendered Lashley's science unnecessary. It answered the questions Lashley himself was addressing, but by completely different means. Lashley was convinced that the solution to the problem of mind lay in biology; Hull argued that biology was not at all crucial to the mind, and the intelligent machine was his proof. By disrupting Lashley's continuity between biology and the mind, Hull's thinking machine also called into question the dependence of intelligence on heredity. If intelligence could be manifested by an entity with neither biology nor heredity, the power of the environment to shape ability suddenly took on a much greater significance. Lashley's science and the assumptions upon which it rested were therefore threatened by Hull's design of intelligent machines.

---

[7] Baernstein and Hull, "A Mechanical Parallel to the Conditioned Reflex," *Science* 70 (1929): 15.

[8] Baernstein and Hull, "A Mechanical Model of the Conditioned Reflex," *Journal of General Psychology* 5 (1931): 99–106.

[9] Krueger and Hull, "An Electro-Chemical Parallel to the Conditioned Reflex," *Journal of General Psychology* 5 (1931): 262–9.

[10] Ellson, "A Mechanical Synthesis of Trial and Error Learning," *Journal of General Psychology* 13 (1935): 212–8. Science Service, "Science at the Century of Progress Exhibition," *Science* 77 (1933): 8–10. I am grateful to Sara Tjossem for bringing the latter article to my attention.

[11] Hull, "The Place of Innate Individual and Species Differences in a Natural-Science Theory of Behavior," *Psychological Review* 52 (1945): 60.

Lashley's opposition to Hull's machine analogy may have been particularly virulent because it drew on an antipathy for such analogies that Lashley had expressed as early as 1924. That year, Lashley argued specifically against the telephone switchboard analogy for the nervous system. A symbol of the reflex theory, the telephone switchboard analogy was widely used to emphasize the automatic character of responses. But for Lashley the analogy was invalid on two counts. The first was that an inflexible system of automatic connections could not account for the plasticity of function which Lashley had observed in the nervous system: equipotentiality, mass action, vicarious function, and temporal variation in function were incompatible for him with the switchboard's rigid system of connections.[12]

The second reason for its invalidity was that the switchboard analogy could not explain the anatomical localization that Lashley believed did exist in the brain. After all, he said, the switchboard need not reproduce the geographical arrangement of the parties it connected; the switchboard of the central exchange could completely scramble those relationships but still function properly. Yet Lashley had found that the cortex preserved the topographic relationships of sensory surfaces, as for example in its representation of the retina.[13] The telephone switchboard analogy was, then, rigid where it should have been flexible, and flexible where it should have been rigid.

Hull himself used the switchboard analogy in his 1943 *Principles of Behavior:* "The condition of organismic need and the status of the environment evoke from specialized receptors neural impulses which are brought to bear jointly on the motor organs by the central ganglia of the nervous system acting as an automatic switchboard."[14]

During the debate with Hull, Lashley broadened his attack on analogies to all kinds of machines. By the end of his career he was arguing not only against the use of analogy in science but against the use of theory itself. The contrast to Hull could not be more striking: wedded to theory all his life, Hull depended on the analogy between organisms and machines as an integral part of his system.

By 1951, at the end of his debate with Hull, Lashley was distancing himself from all kinds of mechanical metaphors for the mind:

> I am less impressed with the analogies between the various machines and neural activity, such as are discussed in *Cybernetics*. . . . Descartes was impressed by the hydraulic figures in the royal gardens and developed a hydraulic theory of the action of the brain. We have since had telephone

---

[12] Lashley, "Studies of Cerebral Function in Learning VI: The Theory that Synaptic Resistance Is Reduced by the Passage of the Nerve Impulse," *Neuropsychology of Lashley,* 140.

[13] Lashley, "The Mechanism of Vision VIII: The Projection of the Retina Upon the Cerebral Cortex of the Rat," *Neuropsychology of Lashley,* 268, and "Functional Determinants of Cerebral Localization," *Neuropsychology of Lashley,* 331.

[14] Hull, *Principles of Behavior,* 29.

theories, electrical field theories, and now theories based on the computing machines and automatic rudders.

He expressed succinctly his own antimetaphorical position:

I suggest that we are more likely to find out how the brain works by studying the brain itself and the phenomena of behavior than by indulging in far-fetched physical analogies. The similarities in such comparisons are the product of an oversimplification of the problems of behavior.[15]

In one of his last articles, "Cerebral Organization and Behavior," Lashley again inveighed against machine analogies: "I doubt that any of the [mechanical] models representing brain activities, that we have today, embody the principles of neural integration."[16]

Lashley's opposition to metaphors and analogies included a more general opposition to theory. In his President's address to the American Psychological Association in 1929, for example, he expressed his atheoretical standpoint in a way that was to become typical: "The facts of cerebral physiology are so varied, so diverse, as to suggest that for some of them each theory is true, for all of them every theory is false."[17] In his private correspondence, too, Lashley presented an atheoretical standpoint. To Hull's disciple at the University of Iowa, Kenneth Spence, Lashley wrote: "I just cannot find any convincing evidence in support of any current theory."[18] Spence himself recognized this anti-theoretical aspect of Lashley's work as one of its weakest aspects: "Since Lashley has no theory which will specify when such will happen, he is perfectly safe whatever the outcome."[19]

The Lashley-Hull Correspondence

I see that Lashley is to be at the experimentalists' meeting. I shall accordingly need some protection. I therefore devoutly hope you will be going. Please let us know.

Clark Hull to Kenneth Spence, March 14, 1947[20]

---

[15] Lashley at the American Neurological Association, June 1951, quoted by Stanley Cobb, introduction to *Neuropsychology of Lashley,* xix. For a discussion of cybernetics and its confrontation with psychology, especially Gestalt psychology, see Steve Joshua Heims, "Encounter of Behavioral Sciences with Machine-Organism Analogies in the 1940s," *Journal of the History of the Behavioral Sciences* 11 (1975): 368–373. See also Heims, *Constructing a Social Science for Postwar America: The Cybernetics Group, 1946–1953* (Cambridge: MIT Press, 1991).

[16] Lashley, "Cerebral Organization and Behavior," *Neuropsychology of Lashley,* 539.

[17] Lashley, "Basic Neural Mechanisms in Behavior," *Neuropsychology of Lashley,* 208.

[18] Lashley to Spence, June 13, 1942, Kenneth W. Spence Papers, Box M937, Archives of the History of American Psychology, Bierce Library, University of Akron, Akron, Ohio.

[19] Notes on a "Seminar on Discrimination Learning: Relational (Lashley) and Absolute (Spence) Theories of Discrimination Learning," Spence Papers, M938, AHAP.

[20] Hull to Spence, Spence Papers, M937, AHAP. This was the meeting of the Society of Experimental Psychologists, the same 1947 Princeton meeting at which the Heredity-Environment Symposium was held.

Lashley's correspondence with Hull and Spence represents the culmination of their fifteen-year-long disagreement. During the 1940s and early 1950s the three became embroiled in a bitter controversy over stimulus generalization and the continuity theory of learning which lasted, unresolved, until Hull's death in 1952. The picture that emerges from this correspondence is that of increasing polarization of the two sides: rather than hastening a resolution, the correspondence exacerbated Lashley's differences with Hull and Spence, so that toward the end they were not only disagreeing about the interpretation of the facts, but about what should actually count as the facts. In 1941, at the start of the correspondence, Spence wrote to Hull complaining that he had discussed stimulus-response discrimination theories with Lashley and that Lashley had rejected them without understanding them.[21] By 1950, when the three were still no closer to an agreement, Spence wrote to Hull: "Lashley simply cannot or will not understand what it is we are trying to do . . ."[22]

According to the continuity theory of learning, when an animal is trained to respond to a particular positive stimulus and to avoid a negative one, all aspects of that positive stimulus impinging upon the animal's sensory apparatus are gradually associated with its response. Conversely, all aspects of the negative stimulus associated with the response are gradually weakened by the training. Thus the animal learns in an incremental way, not in an all-or-nothing burst, to respond to all aspects of a stimulus.

The theory of stimulus generalization extended this proposition: when an animal is trained to respond to a particular stimulus, it will also respond to other stimuli according to how similar they are to the original positive stimulus. As Hull put it, stimulus generalization meant that " '[t]he reaction involved in the original conditioning becomes connected with a considerable zone of stimuli other than, but adjacent to, the stimulus conventionally involved in the original conditioning.' "[23]

Lashley objected to stimulus generalization and the continuity theory in his 1946 article on "The Pavlovian Theory of Generalization."[24] He summarized the theory in a way which, he confessed to Hull, "did not stick entirely to the objective plane of criticism:"[25] "The whole mass of excitation coming in from the situation to which the subject is exposed during conditioning is associated, helterskelter, with the reaction."[26] Lashley argued, to the contrary, that the nervous activity immediately organized itself in relation to the stimulus, so that certain elements of the stimulus became dominant and others ineffective. Following the Gestalt theo-

[21] Spence to Hull, January 3, 1941, Spence Papers, M937, AHAP.

[22] Spence to Hull, March 24, 1950, Spence Papers, M937, AHAP.

[23] Quoted by Lashley to Hull, October 1, 1945, Spence Papers, M938, AHAP.

[24] Lashley, "The Pavlovian Theory of Generalization," *Psychological Review* 53 (1946): 72–87.

[25] Lashley to Hull, October 1, 1945, Spence Papers, M938, AHAP.

[26] Lashley, "Pavlovian Theory," 72–3.

rists, Lashley called this determination of the dominant aspects of the stimulus a "set." The nervous system itself determined this set, not any particular features of the stimulus. Only the dominant elements of the stimulus became associated with a particular response, while the others were ignored because the animal was not set to react to them.

There was, therefore, no stimulus generalization: stimuli did not become automatically associated with a response merely because of their similarity to the original positive stimulus. Rather, the organism had to be primed or sensitized to a stimulus to respond to it. And there was also no continuous learning: the aspects of a stimulus that became associated with a response could not be predicted a priori, Lashley wrote, by a blanket law stating that *all* aspects of a positive stimulus gradually become associated with that response. An animal responds not to all aspects of a particular positive stimulus, but only to those aspects that matter for it, that make a difference for it, to which it pays attention. What stimuli an animal chooses to pay attention to, to notice, depends on the animal's set.

Lashley's objection to continuous learning and stimulus generalization were in complete harmony with his emphasis on an innate basis for intelligence. The idea that stimuli, imposed from the outside, fell upon a static system and elicited responses in a mechanical, stereotyped way was unacceptable to him because it denied the role of an innate intelligence that helped direct the organism in its actions. As he had objected to Dunlap's notion that large portions of the brain were quiescent, Lashley was opposed to Hull and Spence's theory that the nervous system was static, waiting to be directed in its activity by the environment. For Lashley the nervous system was an active system, filtering stimuli and organizing itself to respond: "the effective cue is determined by the internal conditions of the organism."[27] The patterns of activity inherent in the brain's biological structure organized responses: the organism was no automatic switchboard operated by the environment.

The correspondence illuminates three issues at the center of this disagreement: how Lashley's work converged on the ideas of the Gestalt school; how the behaviorists charged the Gestalt school with subjective thinking; and how Lashley's hereditarian perspective led him to work on different problems from those that the environmentalists Hull and Spence investigated. Their correspondence also reveals something of the nature of their interactions: Hull, who is said to have disliked debate, tended to let Spence, the more combative personality, do most of the arguing with Lashley. After sparring with Lashley, Spence typically reported back to Hull.

According to Gestalt theory, perception was the act of comparing a figure against a ground. To perceive was not to react to isolated stimuli, but actively to

---

[27] Quoted by Spence in "Selective Learning: Lashley vs. Hull and Spence," undated ms, Spence Papers, M938, AHAP.

compare stimuli, so that one stimulus became dominant while the others became part of the background. The correlations between Gestalt theory and Lashley's thinking are evident: like the Gestalt theorists, Lashley believed that perception was the model for all learning, and that the organization of perception was innate.[28] For Spence and Hull, however, the notion of a set by which the organism could "choose" what to notice and what to ignore was a descent into subjectivity and anthropomorphism. When Lashley freely compared animal and human perception, for example, Spence wrote to Hull: "[Lashley's] anthropomorphic tendencies . . . reveal great differences that can exist between a person who takes an objective and one who takes a subjective mentalistic approach."[29] And Hull replied:

> I am inclined to believe that there are certain kinds of minds which naturally gravitate toward the vague and uncertain. That means there must be a kind of selective process going on which tends to drift the vague and hazy-minded people into the Gestalt camp, and the more logical and tough-minded people into the behavioristic camp.[30]

By 1946, Spence was reporting to Hull that Lashley's statements were the "most irresponsible" he had ever heard from a "so-called scientist," and that Lashley "simply imputes [to animals] all the perceptual responses that human subjects would make when asked to take an introspective set or attitude toward some stimulus complex."[31]

Lashley, however, true to his atheoretical perspective, denied that he was a follower of the Gestalt school and bristled at the suggestion that his science was subjective or unrigorous. "I am, I think, an even more radical mechanist than you," he wrote to Hull in 1945. "My quarrel is not with mechanism as a philosophy but with the specific mechanisms you postulate and which appear to me to involve mechanical principles which are not applicable to the nervous system."[32] Yet despite his disavowal of theory, Lashley was sympathetic to the Gestalt school and associated with its leaders, especially in the minds of Spence and Hull. In 1947, Spence wrote:

> Frankly, now that [Wolfgang] Koehler has been elected to that body [National Academy of Science] I am not very optimistic of ever being approved. The combination of Lashley and Koehler will undoubtedly operate

[28] Lashley, "Experimental Analysis of Instinctive Behavior," *Neuropsychology of Lashley,* 377.
[29] Spence to Hull, June 20, 1942, Spence Papers, M937, AHAP.
[30] Hull to Spence, November 30, 1942, Spence Papers, M937, AHAP. On the response of American psychologists to Gestalt theory, see Michael M. Sokal, "The Gestalt Psychologists in Behaviorist America," *American Historical Review* 89 (1984): 1240–1263; also Sokal, "James McKeen Cattell and American Psychology in the 1920s," *Explorations in the History of Psychology in the United States,* ed. Josef Brozek (Lewisburg: Bucknell University Press, 1983), 273–323.
[31] Spence to Hull, May 28, 1946, Spence Papers, M937, AHAP.
[32] Lashley to Hull, August 14, 1945, Spence Papers, M938, AHAP.

in the same fashion as it did in the Experimentalists. I am firmly convinced that Koehler is a very vindictive person and I know he feels I have been one of the thorns in his side, and I suppose I must plead guilty – although blameless.[33]

At first, Spence and Hull seemed eager to resolve the dispute with Lashley and to get the controversy out of the public forum. In 1942, Spence wrote to Lashley that their "verbal misunderstandings" should be cleared up in private correspondence.[34] And Hull wrote to Lashley in 1945 that "seasoned scientists like ourselves should not take up space in the journals arguing about purely verbal misunderstandings" that they could easily reconcile in private.[35] Lashley, however, replied: "I hope . . . I have not misinterpreted your writings to such an extent as to make these differences merely verbal."[36] And Hull, reading Lashley's "Pavlovian Theory of Generalization," privately admitted to Spence that he no longer expected a reconciliation: "I am strongly tempted to forget our former friendship and let him have what he is asking for."[37]

By 1945, Lashley and Hull could no longer even agree about what should count as the facts. In a remarkable letter dated October 1 of that year, Lashley wrote to Hull as if their disagreement were a matter of their experimental results, but parts of the letter indicate that the facts themselves were in dispute. Lashley implied at first that he and Hull were working on a common problem: "We are trying to get a better understanding of the phenomena of generalization and of stimulus equivalence and I think we can agree as to the facts to be explained."[38]

Their disagreement, Lashley said, was a matter of different definitions of the same phenomenon: stimulus generalization. For Hull, stimulus generalization was an orderly process by which those phenomena most similar to the original stimulus became associated with the reaction; the strength of association tapered off along a gradient as the difference increased between the original stimulus and the subsequent ones. Lashley argued that stimulus generalization was rather a confusion among stimuli on the part of the reacting organism; such confusion might occur, but there would be no definite pattern to it, as Hull predicted. Still, Lashley treated this as a difference in explanation of the same facts. Later in the same letter, Lashley wrote: "We start with the admitted fact that a series of stimuli can frequently be found such that a maximal reaction is obtainable from one, and diminishing reactions from others . . ." How to explain this fact was the issue:

---

[33] Spence to Hull, March 24, 1947, Spence Papers, M937, AHAP. The reference is again to the Society of Experimental Psychologists.

[34] Spence to Hull, May 20, 1942, Spence Papers, M937, AHAP.

[35] Hull to Lashley, September 19, 1945, Spence Papers, M938, AHAP.

[36] Lashley to Hull, October 1, 1945, Spence Papers, M938, AHAP.

[37] Hull to Spence, September 28, 1945, Spence Papers, M937, AHAP.

[38] Lashley to Hull, October 1, 1945, Spence Papers, M938, AHAP.

whether as a result of the animal's innate power of comparison (Lashley), or as a spreading of associative strength (Hull).

But elsewhere in the letter it is clear that the facts themselves were different for Lashley and Hull: Lashley denied that the fact of stimulus generalization existed. It was no longer merely a matter of explaining the same phenomenon slightly differently. "In training animals," Lashley wrote, "one of our most difficult tasks is to devise methods of directing their attention to those aspects of the stimulus to which we wish them to react. As a consequence prediction, except on an empirical, statistical basis, is almost impossible." This meant, Lashley continued, that Hull's stimulus generalization could not exist: "In discussing the summational effects of association with various aspects of the stimulus you have stated that not all aspects are associated with equal strength but seem to imply that all are associated to some extent and summate to determine the total habit strength." This was the phenomenon Lashley denied: "I, on the other hand, have found that few elements or aspects of the stimulus show any evidence of association and that what is associated is unpredictable from animal to animal; apparently the result of chance direction of attention." In other words, Lashley was no longer disagreeing with Hull simply over the interpretation of the same facts; rather, he was denying the existence of certain phenomena which Hull claimed were real. They were disagreeing about what their animals were in fact doing, not just the interpretation of what they were doing.

Lashley and the Determination of Behavior

By the time of his debate with Hull, Lashley's belief in the genetic determination of intelligence had become a strong and consistent theme in his writings. He revived the concept of instinct years after it had been laid to rest by psychologists.[39] As a member of the National Research Council Committee on Human Heredity, Lashley wrote in a 1941 letter that "the problem of mental inheritance is of the greatest social importance. The present trend [in psychology] is toward what I believe to be an over-emphasis on environmental factors in determining behavior."[40]

At the 1947 Symposium on Heredity and Environment, he said:

Discussions of heredity and environment have tended to regard the nervous system, if it is considered at all, as a vaguely remote organ, essentially similar in all individuals and largely moulded by experience. Even the limited evidence at hand, however, shows that individuals start life with

[39] Lashley, "Experimental Analysis of Instinctive Behavior," *Neuropsychology of Lashley,* 372–392. On the demise and revival of the concept of instinct, see Degler, *In Search of Human Nature.*

[40] Lashley to Robert F. Griggs, October 30, 1941, Lashley Papers, Folder "Committee on Human Heredity, National Research Council," Yerkes Regional Primate Research Center, Emory University.

brains differing enormously in structure; unlike in number, size and ar-rangement of neurons as well as in grosser features. The variations in cells and tracts must have functional significance. . . . they cannot be dis-regarded in any consideration of the causes of individual differences in mental traits."[41]

Two years later Lashley reported that the mechanisms of instinctive and intel-ligent behavior seemed "fundamentally the same":

Both are the expression of modes of the perception of relationships, and these modes are genetically determined. Higher levels of intelligence are based on a greater variety of types of organization, but this does not mean that they are any less dependent upon genetic factors.[42]

He concluded the article by noting that "[t]he student of physical evolution must deal with gene variation and the resultant changes in structure. The student of behavior has to consider an additional step: from gene to brain structure, from brain structure to behavior."[43]

Lashley's interest in the genetic basis of behavior, as I have suggested, had its source in his training with Jennings. Far from having shifted his focus away from this early work, Lashley moved into psychology as a way of staking out the geneticists' claim to this area of science. This interest explains Lashley's position in the debate with Hull. Without it, we are left with a puzzle: why was Lashley's dispute with Hull more bitter than his disagreement with Herrick? In their mecha-nistic outlook Lashley and Hull were closer than Lashley was to his associates at Chicago, the Baptist stronghold. Why couldn't Lashley join forces with Hull to beat back the advances of idealism and spiritualism? Instead, he denounced neo-behaviorist theory as simplistic and argued for a notion of brain function that the neo-behaviorists said retreated into fanciful analogies and vain speculation. He held to his notion, I believe, because it was consonant with his training as a geneticist. Lashley's interest was in elucidating the biological basis of difference. In describing his own work on the anatomy of the nervous system, he wrote,

The most important outcome of the work, however, is the discovery of very great individual differences in the size and concentration of nerve cells (variations of from 25 to 100 per cent) localized in various parts of the brain. Such variations, restricted to definite regions, may provide an anatomic basis for individual differences in behavior or capacity. The determination

[41] Lashley, "Structural Variation in the Nervous System in Relation to Behavior," *Psychological Review* 54 (November 1947): 333.
[42] Lashley, "Persistent Problems in the Evolution of Mind," *Neuropsychology of Lashley*, 460.
[43] Ibid., 475.

of their functional significance is a problem of very great importance for understanding the physical basis of special abilities.[44]

The purpose of Lashley's research was to show that psychological functions were controlled by differences in anatomical structure. If biological difference could be elucidated, the differences in ability that they produced (in, for example, mental capacity or sexual function) could also be tested and categorized.

The interest in elucidating biological difference was a recurrent theme throughout Lashley's work, but it is especially clear in some of his late unpublished lectures. For example, in a lecture entitled "Brain and Intellect," Lashley wrote,

> The student of behavior is constantly confronted with differences in the ability of his subjects, whether animals or men; differences between species, differences arising with age, differences between genetic strains. The problem of the physical basis of such differences is far from solved, although it has been a topic of speculation from Galen and Plato to Elliot Smith and Bergson.[45]

This passage was repeated at the beginning of Lashley's James Arthur Lecture, which he delivered at New York's American Museum of Natural History in 1945.[46] "There can be little question, I believe, that the basis of difference must be sought in the central nervous system," he said in a lecture called "Physiological Factors in Intellect."[47]

In the last lecture he ever presented, at the University of Rochester in 1957, he said that this interest in elucidating biological difference was what had turned him against machine analogies for the mind; it explains his distaste for Hull's thinking machines. "Comparison of the nervous system with a digital computer by von Neumann, Wiener and others implies a uniformity in the action of individual nerve cells which must be contrary to fact."[48] Contrary to fact because, for Lashley, the differences in ability which he could so clearly demonstrate by means of tests must have their physical correlates in the brain, the neurons and the genes. "It is not safe to assume that [nerve cells] . . . have the uniformity or individual constancy of such simple structures as telephone wires or vacuum tubes," he warned in his Vanuxem lectures at Princeton.[49] He emphasized individual variability in brain structure, noting that "individual differences in the structure of the same area often exceed the differences between supposedly discrete areas."[50]

---

[44] Lashley, "Seventeenth Annual Report, July 1, 1945–June 30, 1946," Lashley Papers, Yerkes.

[45] Lashley, "Brain and Intellect," (no date), Lashley Papers, Box 2, folder 9, UFG.

[46] Lashley, "James Arthur Lecture," Lashley Papers, Box 2, folder 12, UFG.

[47] Lashley, "Physiological Factors in Intellect," Lashley Papers, Box 2, folder 14, UFG.

[48] Lashley, "Lectures, University of Rochester," Lashley Papers, Box 2, folder 17, UFG.

[49] Lashley, "Vanuxem Lectures," Lashley Papers, Box 3, folder 3, UFG.

[50] Ibid.

That Lashley intended his work on differences in brain structure to address, if not solve, the question of whether genetics or environment more strongly influenced development is clear from another lecture: "The nature-nurture problem has been argued with little conclusive evidence for a century." Lashley's own work would provide the much-needed evidence. He considered differences in brain structure as mediators between genetic differences and behavioral differences; behavior was ultimately caused by the genes. In a letter to Paul Schiller, he wrote,

> I do not see that there is any difference in principle, whether the response is determined by a very restricted stimulus, as in reflexes, or by a less sharply defined one, like a moving object; in each case there is a genetically determined response to a stimulus structure.[51]

For Lashley, the natural progress of a science was from external to internal; from the observation of outward phenomena to elucidation of their inner mechanisms. A science that remained at the external level, Lashley argued, could never provide a real explanation for the phenomena it described. Skinner's behaviorist psychology, for example, restricted its study of learning to the relations between the outward stimulus and response without an understanding of the underlying organic processes. Tolman, Hull, and even Köhler were guilty of keeping psychology at the external level. "My own interest," Lashley wrote, ". . . is in the *internal* phenomena." "The isolation of psychology from physiology is a dangerous procedure. . . . The restriction of the study of learning to the former class [the external reactions] puts it under quite an unnecessary handicap."[52]

In order to illustrate the progress of a science from the external to the internal level, Lashley discussed the development of genetics. Genetics began, he explained, as a science focused on the external expression of traits, but then moved on to study the internal mechanism of heredity in the biochemistry of chromosomes. Ultimately, Lashley believed, all sciences would merge into one; a grand synthetic theory would provide an explanation for all facts, in whatever domain they were discovered. "Fields of science cannot profitably be set off and studied in isolation," he concluded. "There is need for constant effort to synthesize all facts whose relevance may be suspected, and to seek new approaches to the old problems." Psychology and physiology could not be kept apart, and genetics would ultimately converge on their domain. By interpolating this passage on genetics into his discussion of the development of psychology, Lashley reminded his audience that ultimate explanations of behavior would have to be made in terms of the biochemistry of chromosomes.

For Lashley, one of the main criteria for deciding the validity of a theory was its stance on the nature-nurture problem. He saw the debate between the behaviorists

---

[51] Lashley to Paul Schiller, July 19, 1948, Schiller Papers, UFG.
[52] Lashley, "Manuscripts," Lashley Papers, Box 3, folder 14, UFG.

and the Gestaltists entirely in these terms. "The extreme opposites in learning theory," Lashley wrote in an unpublished lecture fragment, "are the reflexology of behaviorism and the holistic interpretations [like Gestalt theory]." While the behaviorists conceived of learned responses as chains of simple reflexes, the Gestalt theorists believed that the structure of the organism itself determined how it would learn, not the incidental association of stimuli falling on a sense organ. The behaviorists saw all learning in terms of the pressure exerted by the environment, while for the Gestalt theorists, the innate structure of the organism was the determining influence. "The structure of the organism is such that patterns of stimuli directly arouse distinctive integrative patterns and the interaction of such patterns obeys laws which cannot be deduced from the principles of association but are inherent in the organic structure." The holistic point of view "directs attention rather to the inherent organization," and for the holistic thinkers the behaviorist emphasis on learning theory "seems rather futile until the innate mechanisms of organization are understood."[53]

While Lashley denied absolute allegiance to one side or the other, he indicated which aspects of each theory he considered correct. He was attracted to the reductionistic emphasis of behaviorism, its conception of complex psychological functions in terms of simple nervous elements. Compared to this, the physiological wanderings of the Gestaltists ("field induction" and the like) seemed to him vague and abstruse. But when it came to choice of problems and to emphasizing innate organization, Lashley's sympathies were with the Gestaltists. The features of behaviorism and Gestalt that he chose to combine were those that supported his hereditarian perspective.

[53] Lashley, unpublished fragment, Lashley Papers, Box 3, folder 14, UFG.

9

Pure Psychology

In his controversies with Herrick and with Hull, we have seen how Lashley maintained a strictly biological approach to psychology, an approach that emphasized the hereditary shaping of behavior and rejected environmental influence. We have seen how he fashioned an image of himself as a "pure" scientist, focused on facts and "basic research," equally unconcerned with theory and with social applications. We have seen how his biological approach and his self-proclaimed neutrality worked hand in hand as he argued against the reformist orientation of Progressive psychobiology and the environmentalism that underlay behaviorism. While both psychobiologists and behaviorists envisioned their science as a means of social engineering, as a tool for the rational management of people, Lashley avoided most of this rhetoric of control. He seems to have believed that the purpose of science should not be social betterment; and the absence of such betterment rhetoric from his work has allowed his neutral image to flourish.

In this chapter, I will show that Lashley's purist, biological standpoint led him not only into the two controversies already discussed, but also into a longstanding opposition to Freudian psychoanalysis. This opposition figured prominently during his tenure at Harvard (1935–1942), and led to the debacle that forced his withdrawal from active membership in the Harvard psychology department. After Harvard, Lashley became director of the Yerkes Laboratories of Primate Biology (1942–1955) where, I will argue, his emphasis on pure science and the biological basis of behavior also helped to shape the program of research.

In the next chapter, I will discuss Lashley's deep-seated political beliefs and their relationship to the biological purism he maintained in his science.

The Harvard Debacle

Lashley's appointment at Harvard was the result of a search for "the best psychologist in the world." For some years, the chairman of the psychology department, Edwin G. Boring, had been trying to hire Lashley away from Chicago to take a position at Harvard. Boring was involved in a battle over the status of psychology, and he believed that Lashley would help him keep it in the realm of pure science. By 1935, Lashley, intent on avoiding administrative responsibilities, had grown wary of his imminent appointment as chairman of Chicago's psychology department, and Boring was finally successful in his bid to secure a position for the neuropsychologist.

By that time, however, Boring's war had intensified, and Lashley's brief tenure at Harvard was marked by a bitter struggle between "pure" biological psychology and psychoanalysis, the latter represented by Henry A. Murray, a physician, psychologist and psychoanalyst, and Gordon Allport, a personality and social psychologist. This struggle, which Lashley's camp lost, led to Lashley's threat to resign and return to Chicago unless he could cease all contact with the department. Boring managed to have him appointed to a research professorship, which carried neither administrative nor teaching responsibilities. In this capacity Lashley remained until 1942, when he took over the directorship of the Yerkes Laboratories in Orange Park, Florida.

The story of the struggle for Harvard psychology between Boring and Lashley on the one hand and Murray on the other has been told many times.[1] In 1926, Morton Prince, a medical doctor and teacher of abnormal psychology at the Harvard Medical School, established the Harvard Psychological Clinic for research into the treatment of psychological disorders. Soon after the clinic opened, Prince hired Henry Murray, also a physician, as his research assistant. When Prince retired in 1928, Murray succeeded him as director and guided the clinic along psychoanalytic lines.[2]

During this period, Harvard's department of psychology, of which the clinic was a part, was undergoing a process of self-definition: Boring had managed to separate it from the department of philosophy and was eager to hire psychologists who would help him define psychology as a science. These appointments included the social psychologist Gordon Allport and Lashley.[3] The struggle between Boring, Lashley, Allport, and Murray which resulted in Lashley's withdrawal centered on the promotion and tenure of Murray in 1936. Boring and Lashley were opposed to the promotion, while Allport favored it. Ultimately, they compromised: Murray was appointed to two five-year terms as associate professor, and the clinic's support was taken over by the Rockefeller Foundation. This compromise, however, did not satisfy Lashley, and for all practical purposes he withdrew from the psychology department.[4]

The episode is important for understanding Lashley because it reveals the depths of his commitment to an ideal of science that he clearly believed psycho-

[1] Rodney G. Triplet, "Harvard Psychology, the Psychological Clinic and Henry A. Murray: A Case Study in the Establishment of Disciplinary Boundaries," *Science at Harvard University: Historical Perspectives*, eds. Clark A. Elliott and Margaret Rossiter (Bethlehem, Pa: Lehigh University Press, 1992), 223–250. See also Forrest G. Robinson, *Love's Story Told: A Life of Henry A. Murray* (Cambridge: Harvard University Press, 1992), especially 144–152, 184–190, 211–228.

[2] Triplet, 226, 228–9.

[3] Ibid., 231.

[4] Ibid., 238–242.

analysis did not fulfill. For Lashley, as we have seen, psychology had to be grounded in the anatomy and physiology of the brain, and in the correlations between brain function and behavior. Only through this physical connection could the science of mind be made sufficiently rigorous or objective. Psychoanalysis, by contrast, entirely ignored the physical basis of mind and focused instead, Lashley contended, on the invention of fanciful stories about people's private obsessions. Neuropsychology, for him, was science, but psychoanalysis was myth: its theories could never be confirmed or disproved by the observation of biological facts.

For Lashley, then, tolerating Murray's clinic with its psychoanalytic tendencies was not simply a matter of including an alternative approach to the study of mind in Harvard's psychology department. It meant, rather, allowing something that plainly was not science to take up space and resources in the department. For Lashley to have approved Murray's promotion and tenure would have meant to compromise the standards of scientific psychology that he was trying to establish. At stake in the Harvard debacle was nothing less that the legitimacy of the field to which Lashley had devoted his career. Our protagonist could not have renounced the principles of his science so easily.

Lashley and Psychoanalysis

In fact, Lashley had long opposed psychoanalysis for its nonbiological, and therefore unscientific, orientation. As early as 1924, he was arguing against the incursions of "Freudism" into psychology and seeking biological alternatives to Freudian theories of sex. That year, he contributed an article to a special issue of the *Psychological Review* dedicated to assessing the contributions of Freud. His article, "A Physiological Analysis of the Libido," was by far the most critical of the four articles in the issue.[5] In it, Lashley claimed that Freud's notion of the libido as a kind of free energy in the nervous system had no grounding in biological fact. Freud's idea that this energy could be stored up – in effect, dammed – and then released was to Lashley patently ridiculous. It reminded him, he said, more of "the behavior of liquids under pressure" than of any physiological processes occurring in the nervous system.[6] Instead, he sought to explain the libido in terms of the reflex responses of behaviorism: a series of conditioned responses to stimuli were responsible for the development of sexual excitation.

Even after he gave up behaviorism and its reflex explanations, Lashley was still not in any sympathy with psychoanalysis. He continued to dismiss it as nonbiological, even while growing increasingly worried that it was usurping the entire field of psychiatry. During the rest of his career he vigorously debated the

[5] Lashley, "Contributions of Freudism to Psychology III: Physiological Analysis of the Libido," *Psychological Review* 31 (1924): 192–202.
[6] Ibid., 194.

worth of psychoanalysis and tried to preserve biological psychiatry as an alternative approach.

Many of his colleagues who were working toward a mechanistic, biological psychology shared Lashley's opposition to psychoanalysis. In fact, this opposition was so taken for granted that psychoanalysis became a laughing matter, a kind of stock joke in the repertoire. In the 1950s Lashley engaged in two correspondences to discuss the merits of psychoanalysis: the first a serious dialogue with a proponent of psychoanalysis, the second a more joking exchange with a neuropsychiatrist sympathetic to Lashley's views.

The correspondence with the psychoanalyst Kenneth Colby focused on Lashley's objections to the psychoanalytic notion of free energy. This energy was supposedly transferable from place to place within the nervous system. Lashley objected that there was no biological justification for such a concept, writing, for example, that "[t]here has never been a serious attempt to evaluate the basic postulates and many ad hoc assumptions of psychoanalysis in relation to neural mechanisms."[7] In response, Colby gave a spirited defense of the idea.

Lashley also argued more generally that psychoanalysis failed to meet the requirements of a good scientific theory. He explained what those requirements were in one particularly notable letter to Colby. According to Lashley, unlike true sciences, psychoanalysis treated people not as individuals, but rather as examples of different types, differentiated according to the supposed sources of psychic energy. A good science, rather than turning its subjects into stereotypes, should emphasize "the need to study each individual as a distinct problem in terms of his own history."[8] He faulted psychoanalysis for lacking predictive value, claiming that psychoanalysts made their predictions "based on simple inference from past parallels." The observations of the psychoanalysts were not theory-neutral: their "many important facts . . . are handled in such a way that it is difficult to cull them from the mass of interpretation." Moreover, he continued, the "elaborate theoretical systems" of psychoanalysis depended on untested assumptions, and its advocates mistook evidence supporting any one of the assumptions "as proof of the whole system."[9]

Lashley also described the faulty methods on which he believed psychoanalysis depended. Its practitioners used the method of striking illustrations, as when a coincidence is observed and a causal relationship inferred; the method of selected coincidence, in which observations are made until one is found that confirms the theory; the principle of liberality, the opposite of simplicity; the Janus principle, in which facts contradicting a theory are used to claim that "the same cause can produce opposite effects"; and validation of the theory by therapeutic suc-

---

[7] Lashley to Colby, January 28, 1957, Lashley Papers, Box 1, Folder 2, UFG. This correspondence has been published in *Behavioral Science* 2 (1957): 231–240.

[8] Lashley to Colby, January 28, 1957, Lashley Papers, Box 1, Folder 2, UFG.

[9] Ibid.

cess. This last method was "quite irrelevant," Lashley remarked, as it could be used to prove the scientific validity of everything from "Jesus to Coue, snake oil to Lourdes, Greatrakes to Perkins' tractors, animal magnetism to psychoanalysis. . . ."[10]

Many of Lashley's criticisms of psychoanalysis turn on its lack of objectivity. He told Colby, for example, that

> [y]our suggestion that listening to analyses will convince one of their validity leaves me unmoved. I have heard the same suggestion concerning spiritualistic seances. . . . A similar argument is that one cannot appreciate analytic procedures until he has been analyzed. But analysis is not complete until one has suffered conversion, after which objective judgement becomes difficult.[11]

Lashley's opposition to psychoanalysis, then, was unwavering during the last thirty-five years of his life, and this context explains his adverse reaction to Murray's promotion and tenure. Lashley himself gave an account of his growing disenchantment with the theory:

> I was impressed by it at first (1912–20) and followed the literature rather closely until 1930. Eventually I decided that other forms of science fiction are more rewarding and in the past 25 years have only occasionally sampled the analytic literature, without finding any significant improvement, in method or content, however.[12]

Psychoanalysis fell short of Lashley's scientific standards, mainly because he believed that it ignored biological and neurological evidence.

Lashley's other significant 1950s correspondence about psychoanalysis was with a scientist sympathetic to his point of view. George H. Bishop was a neuropsychiatrist and neurophysiologist at Washington University in St. Louis. He and Lashley treated psychoanalysis not only as unscientific, but as a kind of running joke, unworthy of serious consideration, even as they worried about its increasing popularity. Bishop, for example, wrote to Lashley in 1953, "All psychiatrists do with the smattering of physiology they read in the Ladies Home Journal is to poison an already asphyxiating atmosphere with the most noxious fumes of the most pernicious physiological speculations."[13] And Lashley responded:

> Your comments on psychiatrists warm my heart. Three things are necessary for the successful practice of psychiatry: an impressive personality, helped by a beard, a Harley street address, or a crown of thorns; a mystical theory just above the level of comprehension of the patient to whom it is ex-

---

[10] Ibid.
[11] Ibid.
[12] Ibid.
[13] Bishop to Lashley, Nov. 18, 1953, Lashley Papers, Box 1, folder 1, UFG.

pounded; an elastic conscience. Idiots and intelligent people are not helped by psychotherapy.[14]

In discussing the psychiatrist Lawrence J. Kubie, Lashley pretended to psychoanalyze himself for Bishop's amusement:

> I fear you did not profit as you should from hearing Kubie. Otherwise you would know that no child can possibly ever be interested in anything above the navel, or wonder what is in his mother's mind. The brain, as a squishy object floating in fluid is an obvious symbol of the fetus. When I stick pins in it I am releasing a subconscious aggression against the potential brother who might replace me in maternal affection. Of course, there is a deeper meaning which postal regulations prohibit me from sending through the mails.[15]

Still, Lashley continued, Kubie was among the best of the psychiatrists he had known: "Only within his own specialty does he suffer from 'sympathetic paranoia.' At other times he might be classified as sane."

Lashley's criterion for "sanity" among psychiatrists was the degree to which they adhered to biological principles. Recommending the psychiatrist and neurosurgeon Karl Pribram to a position in the medical school at the University of Michigan, Lashley was careful to distinguish him from the majority of psychiatrists as a biological psychiatrist, not one interested in psychoanalysis. "He has not, thank God, been psychoanalyzed and will therefore approach the subject from a scientific and biological point of view." Pribram, according to Lashley, would help the department develop "the scientific aspects of the subject rather than the popular techniques of clinical practice."[16]

In other instances Lashley mocked psychoanalysis: "Thank Rat! (the God of the American psychologists) that you are not seeking the MORAL ABSOLUTE," he wrote to Bishop.

> I was inoculated with Nietzsche at an early age and immunized against all ethical systems. We at least do not have that form of emotional blocking, though, of course, we are probably trying to replace the Father-God image with the self determination of our own brains and so justify our libidinous strivings.[17]

Lashley's attack on psychoanalysis was a particular instance of his more generalized opposition to any system that tried to bring into psychology principles that he considered religious, ethical, or otherwise value-laden. Lashley's opposition to such principles, which he believed were necessarily unscientific, is familiar from

[14] Lashley to Bishop, Nov. 23, 1953, Lashley Papers, Box 1, folder 1, UFG.
[15] Lashley to Bishop, Dec. 3, 1953, Lashley Papers, Box 1, folder 1, UFG.
[16] Lashley to John W. Bean, Jan. 10, 1958, Lashley Papers, Box 1, folder 14, UFG.
[17] Lashley to Bishop, Jan. 22, 1954, Lashley Papers, Box 1, folder 1, UFG.

his debate with Herrick. Anything reminiscent of a soul or inexplicable on strictly mechanistic principles (according to Lashley's definition of mechanistic), such as the psychoanalytic concept of energy distribution, had to be banished from a scientific psychology.

The psychoanalysts were not the only targets of anti-religious criticism in Lashley's letters to Bishop. He also attacked those psychologists and neurologists who he felt were importing religious ideology into their science: Charles Scott Sherrington, John C. Eccles, and Wilder Penfield. He called Herrick, pejoratively, a "Baptist," and remarked wryly that for all their talk of ethical principles, there were more murderers among the Baptists than in the general population. Even the Gestalt neurologist Kurt Goldstein, to whom he was usually considered sympathetic, he considered a "theologian."[18]

Lashley shared a hatred of psychoanalysis with his second wife, Claire Imredy Schiller Lashley, whose first husband had been the biological psychiatrist Paul Schiller. After Paul Schiller's premature death, Claire Schiller and Lashley worked together to finish and publish *Instinctive Behavior,* the book Paul Schiller had begun. Shortly after Lashley's death, Claire Schiller wrote to Heinrich Klüver,

> Another thing which infuriates me is the eagerness of some to squeeze him [Lashley] into a nice little cozy psychoanalytic schema: the iconoclast par excellence, because he hated his father. . . . I know his quarrel with organized religion and orthodoxy was a deep seated one – even now, he was working on a paper on "The Coming Conflict of Science and Religion" – and that he did not like his father (do you know why, really?) but can't you just see how they'd lap it up, what they'd like to make of it?[19]

The Sex Committee

In 1934, nearly contemporaneous with his appointment at Harvard, Lashley became a member of the National Research Council's Committee for Research in Problems of Sex. This committee had been founded twelve years earlier for the purpose of finding biological alternatives to Freudian theories of human sexual behavior. Established initially under the auspices of the Rockefeller-sponsored Bureau of Social Hygiene, its original membership included the physiologist Walter B. Cannon; Frank Lillie, a reproductive biologist; the psychiatrist Thomas W. Salmon; and Katharine Bement Davis, the director of the Social Hygiene

[18] Although Lashley shared an abhorrence of religion with the behaviorists, they too came in for criticism in these letters. Lashley characterized Hull's neo-behaviorism as "ludicrous" and "at the pre-phlogiston level of chemistry" (Lashley to Bishop, Aug. 9, 1954, Lashley Papers, Box 1, folder 1, UFG). About B. F. Skinner he wrote Bishop, "I suppose we should be grateful that there are a few who are not, like B. F. Skinner, trying to create a psychology without a brain" (Lashley to Bishop, Jan. 7, 1956, Lashley Papers, Box 1, folder 1, UFG).

[19] Claire Schiller Lashley to Heinrich Klüver, Nov. 30, 1957. [This letter was wrongly dated, and was actually written in November 1958.] Claire Schiller Lashley Papers, "Klüver" folder, UFG.

Bureau. Robert M. Yerkes served as chairman, a position he retained for the committee's first twenty-five years. In 1931, the Rockefeller Foundation took over the direct support of the committee.[20]

Like many other Rockefeller-sponsored efforts, the committee program was centered on basic research rather than applications. As its origins with the Social Hygiene Bureau might suggest, its mandate was to define normality. The implication of its program was that basic, biological sex behavior could be discovered, both in human beings and in other animals, stripped of all environmental and social influences. The identification of this basic behavior would help to define what was normal behavior and what was abnormal. The biological essence would provide a standard for sexual behavior. The committee aimed to uncover this biological standard by approaching the subject of sex from several different angles: not only anatomical and physiological, but also psychological, anthropological, and sociological.

Lashley participated in this committee in a number of ways. As early as 1922, the first year the committee was in operation, he was already receiving grants to carry out research on sensory, motor, and glandular components of sexual behavior in the rat. Lashley's early committee-sponsored work must have impressed Yerkes favorably, since he suggested that Lashley be appointed to prepare an outline of problems in "sex neurobiology and psychobiology" in which the committee might sponsor research. After becoming a member of the committee in 1934, Lashley directed the research of his student Frank Beach on the effects of cortical damage on maternal behavior in the rat, and on sex reversals in the mating pattern of the rat.

The celebratory history of the committee written to commemorate its twenty-fifth anniversary portrays it as having had a liberating effect on sexual mores. The results of the work, it was said, showed that

> in our American community, actual sex behavior . . . is far more varied than
> had been supposed by most people, even by physicians, scientists and

---

[20] On the committee, see Earl F. Zinn, "History, Purpose and Policy of the National Research Council's Committee for Research in Problems of Sex," *Mental Hygiene* 8 (1924): 94–105; Adele E. Clarke, "Money, Sex and Legitimacy at Chicago, circa 1892–1940: Lillie's Center of Reproductive Biology," *Perspectives on Science* 1 (1993): 367–415. See also Vern L. Bullough, "The Rockefellers and Sex Research," *Journal of Sex Research* 21 (1985): 113–125; Bullough, "Katharine Bement Davis, Sex Research and the Rockefeller Foundation," *Bulletin of the History of Medicine* 62 (1988): 74–89; and Glenn E. Bugos, "Managing Cooperative Research and Borderland Science in the National Research Council, 1922–1942," *Historical Studies in the Physical and Biological Sciences* 20 (1989): 1–32. On American sex research more generally, see Clarke, "Research Materials and Reproductive Science in the United States, 1910–1940," *Physiology in the American Context*, ed. Geison, 351–369; Clarke, "Controversy and the Development of Reproductive Sciences," *Social Problems* 37 (1990): 18–37; Clarke, "Embryology and the Rise of American Reproductive Sciences, ca. 1910–1940," *Expansion of American Biology*, eds. Benson, Maienschein, and Rainger, 107–132; and Diana Long, "Physiological Identity of American Sex Researchers Between the Two World Wars," *Physiology in the American Context*, ed. Geison, 263–278.

confessors. The frequency of conduct disapproved under the prevailing moral codes, such as masturbation, homosexuality and extra-marital intercourse, even on the part of persons who hold normal and respectable places in the community, is apparently greater than was estimated by non-scientific conjecture.[21]

Despite this seemingly openminded outlook, the committee sponsored research that would help distinguish normal from abnormal sex behavior. In fact, at the beginning the idea was not only to identify normality, but to enforce it. Frank Lillie, for example, wrote in 1923 that research should be directed toward three main areas, all of which centered on control: the control of sex determination, the control of sexual development, and the control of sexual relationships.[22]

As the program of the committee evolved, however, control received less emphasis in favor of the identification of the biological standard. The study of animal sexual behavior offered an opportunity to uncover basic biological responses without the interference of "emotional and intellectual influences." "To understand its developmental history [of human sexual behavior], its normal patterning, its variations and abnormalities, requires a knowledge of animal behavior in general," Aberle and Corner wrote in 1953. The success of the animal experiments depended on discriminating the standard from the aberrational: "the observer must first know what to look for as typical, normal behavior."[23] For example, the committee supported an "intensive study" of the sexual behavior of the chimpanzee, carried out under Yerkes's direction at the Yale Primate Laboratories. A standard was identified, and the analogy drawn directly to human behavior: "Certain general conclusions have been drawn from all this work on sex behavior in animals bearing upon the question of what is typical and atypical, normal and abnormal in human behavior."[24]

The committee acknowledged that there was a great deal of variation and modifiability in sexual behavior among primates, especially among humans, but their belief was unshaken in a biological standard: "[H]uman sexual behavior is still based upon the fundamental mammalian pattern and to understand it we must recognize that pattern and thereafter seek to learn how it has been changed as a result of factors that peculiarly influence human beings." These factors, of course, were nonbiological: the various elements of society and environment "which may obscure the basic behavior cycle by overriding the hormonally conditioned sexual responses."[25]

---

[21] Sophie D. Aberle and George W. Corner, *Twenty-Five Years of Sex Research: History of the National Research Council Committee for Research in Problems of Sex, 1922–1947* (Philadelphia: W. B. Saunders Co., 1953), 49–50.

[22] Ibid., 17.

[23] Ibid., 51.

[24] Ibid., 54.

[25] Ibid., 55, 54.

The committee also advanced the belief that this basic, biological behavior was genetically determined, a conclusion drawn directly from the work with animals:

> the pattern of neuromuscular coordination that operates in copulatory behavior is organized very early in life. Somehow it is a part of the gene-determined functional equipment of the individual, and is ready to go into action when the sex-gland hormones enter the picture.[26]

Pervading the committee's work was the notion that biology was fundamental to the human sciences and that it could define the normal or natural type of behavior upon which psychological, social, or cultural influences acted.

One interesting result of this emphasis on a basic, standard mammalian pattern of sexual behavior was the notion that the basic pattern was bisexual. Hormonal influence on the basic substrate determined masculine or feminine behavior, just as other environmental influences decided further variations in behavior. "It appears," wrote Aberle and Corner, "that the characteristic pattern of copulatory activity is fundamentally bisexual, the definitely unisexual (i.e., male or female) forms of behavior being called into play by the mating situation."[27] And again: "Sexual activity has been found to have a similar basic pattern in all mammalian species and in both sexes," and knowledge of this basic biological substrate would lead to "an understanding of the typical and the aberrant types of human sex behavior and of factors which condition or control them."[28] In fact, the belief in a biological standard led the committee's researchers to attempt similar studies on humans and animals. Among Lashley's students, such parallel researches became common. To gauge the effect of bodily injury on sexuality, for example, Frank Beach tested maternal behavior in partially decerebrate rats, while Carney Landis worked on the personality and sexuality of the physically handicapped woman.

As a member of the Sex Committee for more than twenty years, Lashley supervised much of the work done under its auspices. In the 1940s and 1950s, its dominant focus was on the research of the sexologist Alfred Kinsey. Lashley was a friend of Kinsey's, and helped him with the physical and psychological aspects of his famous study of sexuality in women. Kinsey is often hailed as a sexual liberator and progressive, encouraging the acceptance as normal and natural of homosexuality and other behaviors previously considered deviant. Lashley, however, clearly admired Kinsey for other reasons: not because his work was progressive, but because it returned the focus of sex research to the biological basis of behavior, where it should be, and undermined the grip psychoanalysis held on the field.

A series of letters that Lashley exchanged with his longtime friend and former colleague John B. Watson shows the esteem in which they held Kinsey. In 1953,

[26] Ibid., 55.
[27] Ibid., 54.
[28] Ibid., 60.

for example, Lashley wrote that he had spent some time with Kinsey, trying to "straighten out" his chapters on physiology, and bemoaned the lack of other biologically oriented researchers in the field: "Our committee has been trying for some years to get other people interested in human sex research but has not been able to find competent people."[29]

After Kinsey's death in 1956, Lashley remarked:

Kinsey's death was a real loss. He was here [Orange Park, Florida] and spent a little time with me in the fall. He did not have very deep insight into problems but he did have drive and enthusiasm to get things done. It leaves no one but the analysts in the field.[30]

Watson responded more as a disappointed follower than as an antagonist:

I wanted to see Kinsey's death. I wanted to see if he couldn't boost the woman volume. It's pretty bad. . . . By the way I understand the analysts are sleeping with their payments [sic] – probably does them a lot more good than all the verbal analysis.[31]

Lashley immediately rushed to Kinsey's defense:

I think you are a little hard on Kinsey, wishing him dead. He was a difficult person and the female volume is certainly bad. He offered me $1,000 as a consultant on it for physiology and psychology. Since I was a member of the sex committee, I couldn't take the pay, but I spent two weeks in Bloomington, working 10 hours a day on the manuscript. I managed to correct his grammar but could do nothing with him on questions of organization and interpretation.

Nevertheless, Lashley continued,

After 20 years on the Sex Committee . . . I consider Kinsey a Treasure. With plenty of money it is still impossible to get anyone to do decent work on human sex problems. Psychologists who want to use projective techniques and analysts who want their fees paid are a dime a dozen but in the last 10 years not one intelligent project has been submitted.[32]

And Watson concurred:

How could my wrist bone become so disconnected as to write anything that made you think I wanted to hasten the demise of Kinsey. I am sure your estimate of him is the same as mine. To undertake such a work – the endless

[29] Lashley to Watson, May 1, 1953, Lashley Papers, Box 2, Folder 3, UFG.
[30] Lashley to Watson, October 18, 1956, Lashley Papers, Box 2, Folder 3, UFG.
[31] John B. Watson to Lashley, undated, probably 10/56 or 11/56, Lashley Papers, Box 2, Folder 3, UFG. By "payments," Watson meant "patients."
[32] Lashley to Watson, November 12, 1956, Lashley Papers, Box 2, Folder 3, UFG.

drudgery of it. My conscience hurt me a lot. He wanted to see me several times but I could never spare the time. I think I could have helped him a lot on the woman angle.[33]

From his negative experience at Harvard, his work for the Sex Committee, and his encouragement of Kinsey, we can see that Lashley's interest in the biological basis of behavior underlay his fierce opposition to psychoanalysis. But in Lashley's view, psychoanalysis was doubly damned. Not only were its theories innocent of biology; its clinical applicability offered premature promises of a quick fix to psychological problems, before scientific research, unconcerned with practical outcomes, had uncovered the laws of mind, brain, and behavior. To counter this corrosive effect of psychoanalysis on the science of psychology, the Sex Committee during Lashley's tenure distanced itself from the practical control of behavior and moved instead toward an understanding of its biological basis. Pure research, not applications, was the focus of the committee's work.

Lashley's complete dismissal of psychoanalysis is best summed up in another of his letters to Watson. What a pity, he wrote his friend in the winter of 1956,

> that [Adolf] Meyer [the psychobiologist] did not influence more people. I got pretty well acquainted with him in later years, serving on the sex committee with him, but he was always vague and negativistic. He was, however, sensible about Freudian psychology – called it a pretty, Jewish, fairy tale.[34]

The Yerkes Laboratories

Lashley left Harvard in 1942 to become director of the Yerkes Laboratories of Primate Biology in Orange Park, Florida, a position he held until his retirement in 1955. His directorship illuminates in greater detail the same themes evident in his ordeal at Harvard and in his membership on the Sex Committee: his drive to construct a pure, biological psychology. At Orange Park, Lashley guided the work of the laboratories away from his predecessor Yerkes's social engineering emphasis and toward an investigation of the biological – or, more precisely, genetic – basis of two major types of behavior, sexual and intelligent.

The labs were organized as the Yale Laboratories of Primate Biology at Yale in 1925 by Robert M. Yerkes in order to "investigate the . . . usefulness of breeding anthropoid apes" for scientific purposes. In 1930 a breeding station was established at Orange Park because the northern Florida climate was better for the apes, and the following year a new laboratory was constructed at Yale. For the next twelve years research was done both at Yale and at Orange Park. In 1942, upon

---

[33] Watson to Lashley, n.d. (probably 11/56 or 12/56), Lashley Papers, Box 2, Folder 3, UFG.
[34] Lashley to Watson, January 1, 1956, Lashley Papers, Box 2, Folder 3, UFG.

Yerkes's retirement, the name of the lab was changed to the Yerkes Laboratories, and Lashley took over the directorship.[35]

From the beginning the lab had trouble finding sources of funding. The Rockefeller Foundation and the Carnegie Corporation were its steadiest supporters, but their continued support required constant bargaining by Yerkes and Lashley. Yale and Harvard were harder to commit to a long-term investment in the lab. In 1938, for example, Yerkes wrote in frustration to the dean of the Yale Medical School,

> This undertaking is being penalized to an extent which it would be difficult to exaggerate. It must be perfectly plain to anyone who has even casual acquaintance with the laboratories and their work that the present uncertainties, and only slightly less the inadequacies of plant and support, render progress extremely difficult.[36]

Although Yale had at first mediated the relationship between the lab and the Rockefeller Foundation, it withdrew support entirely in 1952 because the lab was not making a sufficient contribution to teaching and research at the university. Harvard, which had contributed only Lashley's salary, withdrew that support in 1955 after his retirement. During the Second World War the funding problem was acute, and there was constant talk of reorienting the work of the lab toward war work, though little evidence exists that this was ever carried through.

Like Lashley, Yerkes believed that the study of chimpanzees would provide a clue to human nature, because in chimps the basic biological reactions were "less obscured [than in humans] by cultural influences."[37] But in contrast to Lashley's purity, Yerkes's social engineering outlook pervaded the work of the laboratories. The research there, he argued, would affirm the truth of his ringing declaration: "[p]robably there is no more unprofitable assumption than that of the unalterability of human nature."[38] For Yerkes, the breeding of an ideal laboratory chimpanzee would serve as a model for the same shaping of human beings to specification. In the sixth annual report of the laboratories for 1934–35, Yerkes described the kind of research animal he was trying to create.

> The ideal experimental chimpanzee should then be small, with tough skin, thick heavy coat, long and delicate hand and foot; vigorous, robust, assimilatively efficient, disease-resistant, preferably with specific immunities; short gestation period, rapid maturation and development, high fertil-

---

[35] On the Yale Laboratories under Yerkes, see Donna Haraway, *Primate Visions: Gender, Race and Nature in the World of Modern Science* (New York: Routledge, 1989), 59–83.

[36] Robert M. Yerkes to Dean S. Bayne-Jones, May 20, 1938, Lashley Papers, Folder "Committee on Yale Laboratories," Yerkes.

[37] Yerkes, *Chimpanzees: A Laboratory Colony* (Yale University Press, 1943), 3.

[38] Ibid., 9. On Yerkes as a social engineer, see Donna Haraway, "The Biological Enterprise: Sex, Mind and Profit from Human Engineering to Sociobiology," *Simians, Cyborgs and Women: The Reinvention of Nature* (New York: Routledge, 1991), 43–68.

ity, twin-producing; behaviorally highly adaptive, active, original, non-destructive, coöperative; naturally tame and readily gentled, non-pugnacious, affectively stable and with high emotional threshold, unselfish or altruistic, frank, dependable, easy to handle, good-natured, even-tempered; and finally – a far cry – as nearly as practicable homozygous.[39]

These criteria are both physical and mental; indeed, Yerkes directed the investigations equally toward the biological and psychological aspects of his research subjects. While Lashley argued that the principal justification for an anthropoid lab came from the study of brain and intelligence, Yerkes was anxious not to have the lab specialize too early in chimpanzee psychology. He was concerned that the work of the labs be not "exclusively psychobiological," but "more broadly biological." Yerkes wrote, for example, that the lab's research programs "were from the beginning inclusively biological instead of merely psychobiological," even though he considered himself a psychobiologist.

In the prologue to his 1943 book *Chimpanzees: A Laboratory Colony*, Yerkes explained his eugenic ideal. The breeding of the chimpanzee, the "servant of science," could be directly applied to the human situation:

> We have believed it important to convert the animal into as nearly ideal a subject for biological research as is practicable. And with this intent has been associated the hope that eventual success might serve as an effective demonstration of the possibility of re-creating man himself in the image of a generally acceptable ideal.[40]

Yerkes admitted that agreeing on what constituted such an ideal would be difficult, but then immediately came up with several criteria – all of which, in good eugenic fashion, were mental traits:

> One is led to think at once of generally recognized human shortcomings, such as extreme selfishness, dishonesty, slothfulness, cruelty, and to wonder whether by developing an ape notable for its dependability, trustworthiness, consideration of others, measure of self-control, and coöperativeness, and by exhibiting it extensively and impressively, attention might not be focused on the problem of improving human nature and the possibilities of its solution.[41]

For Yerkes, the servant of science was to be engineered, both physically and mentally, to become not only the ideal research animal, but also to demonstrate the possibility of human engineering. When Lashley took over the directorship of

[39] Yerkes, "Sixth Annual Report, Laboratories of Comparative Psychobiology, Yale University, July 1, 1934 to June 30, 1935," Lashley Papers, Yerkes. Thanks to Donald A. Dewsbury for bringing this passage to my attention.
[40] Yerkes, *Chimpanzees*, 9–10.
[41] Ibid., 10.

the lab in 1942, he reoriented the research program from an emphasis on the general biology of apes and on shaping an ideal research specimen to a focus on the interrelationship of sexual and intelligent behavior.

Lashley scaled back the emphasis on breeding considerably. For example, in the Fourteenth Annual Report, 1942–43, Lashley wrote that L. H. Snyder, a human geneticist and the chairman of the Committee on Human Heredity, spent a week as a guest of the laboratory, studying the breeding program and advising its direction. "His advice is being followed in the breeding program," Lashley noted, but there was no mention of engineering an ideal ape or an ideal human being. Rather, in developing "strains diverse with respect to physical and behavioral traits," Lashley's breeding program would provide evidence of the "genetic factors underlying temperamental differences, and on the inheritance of physical traits."[42] The purpose of the breeding program, then, was not to demonstrate the potential for engineering the ape or the human being, but rather to show that both mental and physical traits were genetically determined. In other words, the aim of Lashley's breeding program was not eugenic, like Yerkes's; it was not supposed to show how radically things could change, but to prove how they must remain wholly the same.

In the Seventeenth Annual Report, Lashley's discussion of the breeding program was even more dismissive than in the 1942–43 report. Again, Snyder's advice was taken about developing diverse strains of chimpanzees, Lashley wrote; but he continued, "[t]his is, of course, a long term project with no great prospect of success but it is worth the small investment of time that it requires."[43]

Under Yerkes's direction, sex and intelligence were, to some extent, the foci of biological and psychological research at the lab. President Charles Seymour of Yale, for example, wrote in 1938 that Yerkes was to be congratulated on his work at the lab showing that the brain and sexual cycle of chimpanzees was very close to that of humans.[44] Under Lashley's guidance, however, sexual and intelligent behavior became the twin poles of the research program.

In his long-term view of the Yerkes Laboratories written in 1950, Lashley described the research plans of the lab, which were centered on seven main topics.[45] First, the researchers studied the evolution of behavior and the origin of human mental traits. Analysis of the psychological functions of the higher apes, Lashley wrote, could determine which of those functions had changed during the course of evolution; physiological and anatomical studies would elucidate the

---

[42] Lashley, "Fourteenth Annual Report, 1942–3," Lashley Papers, Yerkes. Thanks again to Donald A. Dewsbury.

[43] Lashley, "Seventeenth Annual Report, 1945–6," Lashley Papers, Yerkes.

[44] Seymour to Warren Weaver, April 11, 1938, Lashley Papers, Folder "Committee on Yale Labs," Yerkes.

[45] Lashley, "A Long-Term View of the Yerkes Laboratories, 1950," Lashley Papers, Folder "Karl S. Lashley, Second Director," Yerkes.

physical basis of those changes. Second, research was directed toward solving the "perennially recurrent" problem of constitution and environment. In the human case, nature and nurture could not be teased apart: the complexity of the human social environment made isolation of human instincts impossible. But with animals, it was a different matter: "[r]igorous control of the environment of apes is possible and generalization to man largely justified," Lashley wrote.

Third, the researchers focused on experimental psychopathology, to determine the degree to which early experience influenced the origin of neuroses and some "major insanities." Lashley believed that "the persistent effects of early experience are not demonstrated." The fourth topic of research was sex physiology and behavior: comparison between human and chimpanzee menstrual cycles and the influence of "sex secretions" on dominance and aggressiveness. Fifth, the researchers studied "the structure of intelligence" in order to determine and isolate the "independent functions" that comprised intelligence. They used several different methods in this study: the comparative study of the abilities of different animals (Lashley's student Hebb, in fact, designed a Binet intelligence test for chimpanzees); factor analysis of individual differences using Thurstone's methods; and the disintegration of behavior by brain injuries.

Sixth, Lashley proposed that his researchers study brain structure and function by examining the symptoms of and course of recovery from localized brain lesions in the animals. Current studies, he noted, were leading to a mass-action interpretation of brain function, in which clinical symptoms were the result of the "malfunctioning of injured tissue rather than of loss of brain tissue." Finally, the research of the lab was directed toward the study of certain medical problems. In the past these had included infantile paralysis and dysentery. But current studies reflected Lashley's interests in brain function and vision: the effects of electroshock on behavior and brain tissue, and of light privation on visual defects.

These seven topics show clearly that Lashley was directing research toward intelligence – determination of its biological basis and the role of nature in shaping it – and on sex. In the lab's research program, sex and intelligence were parallel kinds of behavior in that both had a strong biological component which could be isolated by holding all environmental influences constant. While Lashley admitted that in human behavior constitutional and environmental factors were inextricably intertwined, he nevertheless believed that the two were ultimately separable influences. His idea was to show how isolable they actually were, not how completely they were involved with each other. The fact that he believed that environment could be held constant while natural endowment was investigated shows where his interests lay. Lashley was studying apes in order to excavate the "natural" essence that dwelt beneath the surface of every socialized human being. Drawing a parallel between sex and intelligence was simply a way of elucidating those natural tendencies. Like sexual behavior, intelligent behavior depended not on culture but primarily on brain structure and "secretions." Although it was perhaps the highest psychological function, intelligence was firmly grounded in nature. Moreover, sex and intelligence were mutually interactive functions: sexual "secretions" had an influence on the "dominance" and "aggressiveness" of an animal, and puberty effected "personality" changes in chimpanzees. Brain damage, in turn, affected sexual behavior.

Under Lashley's direction, the lab moved toward research into the biological basis, the constitutional component, of these complex behaviors. For Lashley, the function of a primate laboratory was "the differentiation of innate and acquired characters." He directed the research of the lab toward the fulfillment of this specific goal and away from its former general biological studies and breeding experiments.

In the last appointments of Lashley's career, then, we can see how he continued to construct and to fight for his scientific ideal. In the psychological science he tried to promote at Harvard, on the Sex Committee, and at the Yerkes Laboratories, theory did not stand in the way of facts; his science (as he explained to Colby) was built on an objective method and on value-neutral observations. Lashley's ideal science was also free from the constraints of immediate practical applicability; not the healing of mental ills nor social engineering was its goal, but the pure understanding of nature's truths. Ultimately, for Lashley, those truths rested in biology or, more precisely, in a rather restricted sector of biology: genetics. His ideal science did not simply locate the source of mind in the biology of the brain, but screened out environmental and cultural influences and showed how deeply both animal behavior and human action were genetically determined.

10

Public Science and Private Life

Prologue

To this point, I have maintained that the distinguishing features of Lashley's science were his opposition to theory and applications and his rhetoric of neutrality, and that these were accompanied by his increasing conviction that intelligence was biologically based and hereditarily determined. These last two chapters will demonstrate that Lashley's trademarks were more than simply coincidental aspects of his neuropsychology. Taken together and placed in their historical context, these characteristic features of Lashley's science comprised a powerful political statement: they made an argument for the social status quo, and against political change, progressive reform, and particularly against racial integration.

How can this be, though, when statements on race, or political views of whatever sort, are nowhere to be found in Lashley's work; when, in fact, his deliberate neutrality precluded any reference of the kind? Why would a scientist cultivate such a studied political neutrality, if his purpose were to address the political problems of his age that troubled him? And how can a historian presume to know what those political problems were, and whether or not the scientist's work meant to address them, if there is no mention of politics of any kind anywhere in the body of that work?

It is true, as we have seen, that there are no explicit political references in Lashley's scientific papers: his public persona seems entirely apolitical. But in the unpublished record lurks quite another story. Lashley's private self – the face he turned to friends in correspondence – differed wholly from his public stance. I am not referring now to his scientific correspondence – such as the letters he exchanged with Herrick, Hull and Colby – but to letters he wrote to his close friends. Any comparison of Lashley's published works with these unpublished letters leaves the reader with an acute sense of cognitive dissonance. The private letters present an image of Lashley that seems utterly inconsistent with the impression left by his scientific papers. On the one hand, there is Lashley the neutral scientist, renowned for his meticulous experiments and devotion to the facts; and on the other, there is Lashley the private man, who expressed with vehemence his views on race and on the social order.

What is the historian to make of such an incoherent picture? The most obvious and easiest thing to do is to deny any relationship between the public science and the private letters. Lashley made certain comments privately, but clearly they never intruded onto his scientific work: evidently he was just an inconsistent person. On this reading, we could simply acknowledge the existence of the letters, admit that many people during the first half of this century shared Lashley's opinions, and move on. A scientist's private life should, after all, be irrelevant to an understanding of his science; prying into that private life is the kind of thing only an antiquarian (at best) or a voyeur (at worst) would be interested in. Lashley's private writings are a historical curiosity, but of no significance in understanding his work, which can be understood and appreciated without this glimpse into the personality of the man behind the science. In this interpretation there seems not to have been – indeed there could not have been – any relationship between his public and private sides.

But there is another answer to the Lashley problem, one that starts from the assumption that science, whatever else it may do and may be for, no matter how value-neutral and factual it may seem, is always a political tool. Science is a human activity, and so even while it reflects nature and is constrained by the "real" world, it is also necessarily molded by the concerns of the people who practice it: it fulfills their visions, and it answers their needs. As Steven Shapin and Simon Schaffer have said in *Leviathan and the Air-Pump:*

> We shall suggest that solutions to the problem of knowledge are embedded within practical solutions to the problem of social order, and that different practical solutions to the problem of social order encapsulate contrasting practical solutions to the problem of knowledge.[1]

I suggest that Lashley's solution to the problem of brain function and behavior in rats was simultaneously a solution to the problem of social order in America.

I do not mean to imply that science is in some sense "driven" by social interests; that the politics are prior to and determinative of the scientific work. This model is far too reductive, and it leaves us with "the political" or "the social," an unanalyzed residue that seems to require no further explanation. What determines the political or social interests? The social interest theorists either do not ask this question, or do not deem it worthy of an answer.

Nor should science be treated as a kind of main event taking place against a vague cultural background or Zeitgeist. The historian is then relegated to searching for broad analogies or resonances between the science and the culture. Can Lashley's equipotentiality theory somehow be seen as correlated with the mid twentieth-century rise and spread of Marxism?[2] Where the social interest model is

---

[1] Steven Shapin and Simon Schaffer, *Leviathan and the Air-Pump: Hobbes, Boyle and the Experimental Life* (Princeton: Princeton University Press, 1985), 15.

[2] See, for example, Harrington, *Medicine, Mind and the Double Brain*, 270–1.

overly reductive and deterministic, the Zeitgeist model is just too vague. Instead, I want to suggest a model in which science and politics are on an equal footing, in which they develop together, and as a result are always complexly intertwined.

In Lashley's case, the great value of his private correspondence is that it points to the political context relevant to a historical interpretation of his science. From this correspondence, most of it dating from the 1950s, we can see that ideas about race and fears of race mixing were of great concern to him.

The correspondence shows that racial integration was more than a general Zeitgeist against which Lashley did his scientific work. Rather, he was genuinely preoccupied with the movement of African Americans into white society, and expressed this preoccupation again and again in letters to friends. He probably would not have dared to air his views publicly, especially in the atmosphere of antiracism that had enveloped the social and biological sciences by the 1950s. But, even given this atmosphere, it is notable that Lashley's private beliefs were not considered egregiously inappropriate. None of his correspondents disagreed with him; in fact, in their own ways, they confirmed his perspective. This suggests that concern over the negative effects of race mixing was part of a powerful set of beliefs that was aired privately among a certain scientific elite, but remained tacit and unacknowledged in public.

In order to portray the past fairly, we cannot afford to dismiss Lashley's private correspondence, and lose this rare glimpse of what was going on behind the scenes in the mid twentieth-century atmosphere of depoliticization and antiracism. I will suggest, using the correspondence as evidence, that Lashley's private views on race were in dialogue with his public science, that they informed it and were informed by it, and that they helped to justify it. I will argue that only in the light of the correspondence, and the political views expressed therein, can we make sense of the central features of Lashley's science – the neutrality, the oppositional stance, the hereditarianism – and see how they are integrally and necessarily interrelated.

In his letters, Lashley was an impassioned advocate of the status quo, a dedicated opponent of any change to the social order he had always known. In his science, the same beliefs appear, but in a different guise. His desire to remain relentlessly neutral stemmed not so much from a devotion to and a love of fact (though these he certainly possessed), as from his conviction that scientists should keep their hands off society, disclaiming any social reformist or ameliorative impulses. His hereditarianism dovetailed neatly with this belief: science should not try to change society, because society could not really be changed. People could not alter their native endowment; how they were born was how they had to stay; the social order must remain as it was. Hereditarianism proved, Lashley believed, that racial integration was a lost cause; and so, to keep his sights fixed on the "facts" of brain function was, in effect, to work against this or any other form of social change.

Lashley's Private Politics

Most of Lashley's surviving private correspondence (and again, I am speaking here of letters to friends, not his scientific correspondence) dates from the 1950s. What happened to his earlier letters to friends is anyone's guess. Perhaps he did not keep copies of them, as he did with his later letters. Or perhaps, if he did make copies, these were lost or inadvertently destroyed during one of his many relocations during the 1920s, 1930s, and 1940s. Or perhaps, as an intensely private man, Lashley destroyed the letters himself. In any case, I have not been able to locate them. However, even such relatively late correspondence can provide us with important clues about Lashley's beliefs in his younger days. There are good reasons to suppose that the opinions he expressed in these letters had been with him throughout his life, that his private persona remained remarkably consistent over the years.

One hint of such consistency comes from the autobiographical interview that Lashley granted Anne Roe in 1950. He described for her how, nearly forty years earlier, as a laboratory assistant at the University of Pittsburgh, he had set up a segregated classroom.

> I had had a good bit of race feeling. There were Negroes in the town at home [Davis, West Virginia]. I don't recall feeing it then. . . . In Morgantown [West Virginia, where Lashley attended college] there were no Negroes allowed and I approved. I got to Pittsburgh and there were these three big, buck niggers in the class. I went out and paced up and down the hall, debating what I was going to do, whether I was going to give up and go away right then. But, finally, I came back in and I put the Negroes in a corner at one table and put another table between them and everybody else and then I arranged the others as far away as possible. There were no protests. I got away with it. . . . This prejudice to which I still adhere was very prominent at that time.[3]

Lashley's story is not only a testament to his actions in 1912; it is also an eloquent expression of his state of mind in 1950. He reimagined and reexperienced the scene not in order to disavow his racism, but to underscore the fact that it was still with him. "This prejudice to which I still adhere" was not merely mentioned in passing, but given life by recounting a dramatic episode that Lashley could still envision nearly forty years later. The point is not how distant the episode was or how foreign the feeling, but how much a part of his imagination it still continued to be.

Of Lashley's late correspondence with friends, several exchanges survive, including those with the neuropsychiatrist George Bishop, the immunologist

---

[3] Transcript of an interview of Lashley by Anne Roe, 1950, Anne Roe Papers, Collection B:R621, American Philosophical Society Library.

William Taliaferro, and the comparative psychologist Heinrich Klüver. In all of them Lashley expressed similar views on the "Negro problem." For example, in a 1956 letter to Bishop, Lashley wrote in his accustomed oppositional style: "The frustrations of neurophysiology are disheartening. As I look back over my 40 years of work, all I have accomplished is to throw doubt on every theory of cerebral integration, including my own." Lashley said that the neo-behaviorist B.F. Skinner was "trying to create a psychology without a brain" and that his own sympathies were with physiological psychology. In this brief letter are all the usual signs of Lashley: the disavowal of theory, the emphasis on physiology of the brain, the opposition to behaviorism. The letter ends with the following passage: "I have just been to Jamaica, in the hope that it might be a pleasant place of retirement. Wonderful climate and beautiful scenery, if you cannot tell black from white. But I still prefer segregation."[4]

Just as Lashley's antibehaviorist comments were typical of him, so a remark in favor of segregation was not at all extraordinary; in fact, this comment is milder than the usual racist remark. In a 1955 letter after the same Jamaican excursion Lashley wrote to Taliaferro,

> It was my first experience in a black country. By the time I got past the immigration officers I was ready to hold my own private lynching. The island is beautiful but not the population. Heil Hitler and Apartheit![5]

In a letter to Klüver, Lashley again related his experiences in Jamaica:

> There are some able young English teachers in the medical school [in the University of Jamaica] but the students are all black. In fact, during a week in Kingston I did not see more than 20 white faces. Too bad that the beautiful tropical countries are all populated by negroes. Heil Hitler and Apartheit![6]

By far the most remarkable of Lashley's exchanges in his last decade was with his old friend and former Hopkins colleague, the behaviorist John B. Watson. Of all his late correspondences, the one with Watson was the most sustained, beginning in 1951 and ending only with the deaths of the two principals in 1958. It was the most intense of Lashley's private exchanges, containing over thirty letters in all. And it is the correspondence that gives the best sense of the private world that Lashley inhabited, and that he shared with at least some of his scientific cohort. Their letters show that, though they expressed it in different ways, Lashley and Watson were in fundamental agreement about how society should be ordered.

In 1951, at the age of 61, Lashley was director of the Yerkes Laboratories in Orange Park, Florida, and into his third year as a widower (his wife, Edith Baker

[4] Lashley to Bishop, Jan. 7, 1956, Lashley Papers, Manuscript Collection 89, Box 1, folder 1, UFG.
[5] Lashley to Taliaferro, Dec. 22, 1955, Lashley Papers, Box 1, folder 12, UFG.
[6] Lashley to Klüver, Jan 2, 1956, Claire Schiller Lashley Papers, Folder "Klüver," UFG.

Lashley, had died in 1948). Watson, 73, was living in retirement on a farm in Connecticut, having withdrawn into seclusion some fifteen years earlier following the sudden death of his beloved wife, Rosalie Rayner. His only occasional companion seems to have been his New York City-based secretary, Ruth Lieb. Lashley and Watson had been out of touch for quite some time; why exactly they sought each other out at this point and revived the friendship is not clear. In any case, they were able to forget all their earlier scientific disagreements (which must have seemed very far indeed in the past) and pick up the friendship where they had left off. The first letter of the series was Watson's, dating probably from December 1951, in which he expressed belated condolences for the death of Lashley's wife, and confessed his eagerness for a real confidant: "So glad we can drop the middle aged conventions. I would love to see you and have a good long session on what psychology ain't. . . . Come up soon."[7]

Lashley responded, explaining his own living situation in a tone equally eager:

> After Edith died, I built a small cabin on the river here [Jacksonville] and have been batching; prefer it to putting up with a housekeeper or the unreliable nigger servants here. I have become an expert on piecrust and have lived chiefly on mince pie this winter; still keep the weight down to 126. I wish you would break away from that damned New England climate and spend some time with me here. I can also cook good steaks and am a fair bartender.[8]

The two friends finally did meet again at Watson's farm in late 1952, after which their letters became increasingly intimate. Over the next several years, the two shared their most private concerns with each other. For Lashley, as I have already suggested, this private preoccupation was racial hierarchy and race mixing; Watson's, as it emerges from the exchange, was sex. I will discuss their correspondence in roughly chronological order, showing that it falls into two parts. Watson's concern dominated the first part, but was eclipsed in the second part by Lashley's preoccupation after his 1955 visit to Jamaica. I will argue that the two men's obsessions, sex and race, were complementary halves of a single coherent worldview.

Renewing his contact with Lashley helped to revive Watson's interest in psychology after his long isolation from the field. But Lashley was not only Watson's link to the scientific world; he was also a close male friend with whom he could share his most private fantasies. Indeed, neither Watson's letters nor Lashley's draw a sharp distinction between discussions of scientific matters and of personal experience: in this correspondence, public and private completely overlapped. Sex was of central concern to them not only as a private matter but as an object of

---

[7] Watson to Lashley, n.d., probably 12/51, Lashley Papers, Box 2, folder 3, UFG.

[8] Lashley to Watson, April 6, 1952, Lashley Papers, Box 2, folder 3, UFG.

scientific inquiry. It is unclear in the following passage, for example, whether Watson was writing about his own experience, or a scientific experiment, or both.

> I wish you and Kinsey and I could have a few sessions. I wonder if any correlations exist between clitoris location and entrance to vagina. I should think the nearer the entrance the better the chance for the male to stimulate it during coition. Also whether a woman who experiences an orgasm regularly (if such exists) can teach another woman? I have made considerable study of a lot of things in the field, so to speak, and I grow more modest after each research![9]

Lashley commiserated with Watson that as female sexuality was a mystery to men, it was also a mystery to science:

> Since I saw you I have spent another week with Kinsey trying to straighten out his chapter on physiology. The committee on sex research of the NRC has been supporting some work on the sensory endings of the female genital tract; the labia minor seem to be as well supplied [with nerves] as the clitoris, but of course we do not know the function of the different nerve endings. Kinsey has a chapter on the techniques of masturbation in women; stimulation of the labia minor seems as frequent as of the clitoris. He has a lot of material on orgasm but none of it seems to further the problem of frigidity.[10]

As these comments indicate, the problem of sexuality for both scientists was essentially the problem of women's sexuality, especially frigidity and the ever-elusive female orgasm. Their scientific perspective on these "female problems" depended on objectification and control. Women had to be objectified so that they could be studied scientifically, and the purpose of such study was to control their behavior: to increase what was lacking and to decrease what was superabundant. But their letters made it clear that objectification in the scientific sense, and control through physiological, psychological, and anthropological research, were continuous with objectification and control in their more private aspects. The treatment of women as objects of scientific study resembled their treatment as objects of sexual fantasy and longing. Likewise, the control that science brought to sexual matters can easily be understood as an extension of the desire for such control in the private realm.

The two friends' semiprivate, semiscientific discussions of sex were often accompanied by a good deal of boasting and bravado. Explaining to Watson that he had had to decline an invitation to lecture on the psychology of sex, Lashley remarked, "My nice nurse is coming again twice a week to give me shots of testosterone but she occasionally brings her husband with her – a huge ex-Marine

[9] Watson to Lashley, April 17, 1953, Lashley Papers, Box 2, folder 3, UFG.
[10] Lashley to Watson, May 1, 1953, Lashley Papers, Box 2, folder 3, UFG.

who does not encourage trespassing."[11] When Lashley used some of his "excess income" to set up a fellowship fund in neurobiology, he told Watson that getting rid of the money would be a guard against "woman trouble; some insurance against senile infatuation."[12] When Watson heard this, he chastised his friend (more than once) for not using the money to set up a "streamlined Western harem," which Watson said was his favored method of consorting with women. (Lashley seems to have preferred a dog: "less talkative and less final.")[13] Lashley reported that at a party he had drunk four tumblers of straight whiskey and "did not lose enough inhibitions even to flirt with the other men's wives there."[14] Watson replied he did not understand how Lashley could have had "all that booze . . . and not rape somebody in the room. . . . I should think that the drinks would at least have started you singing the S. Pacific song 'I want a dame.' By the way have you had a check up recently on the old prostate?"[15]

Concern for their own sexual potency led Lashley and Watson to have a very direct interest in contemporary work in endocrinology, especially the various experiments by Eugen Steinach and Paul Kammerer that were supposed to increase sexual function. "Several of my friends," Watson reported to Lashley in 1955, "have had their vas tied back and claim increased sexual stirrings. Any truth in this? The removal of the fear of knocking a gal up might account for it."[16] Lashley replied that Steinach's idea of tying back one of the vas deferens, of which Kammerer was an enthusiastic proponent, had been discredited, and any effect was probably due to "local inflammation." Lashley used the work at the Yerkes Lab to dissuade Watson from trying it: "[c]urrent work here indicates that sex motivation in the chimp (and so probably in man) is nearly independent of hormone action. . . ."[17]

From their letters, it is also clear that Watson and Lashley thought that the sexuality of the women of their own society differed from that of the "primitive" women of countries like Africa and China. For Western white women, the problem (according to Watson and Lashley) was frigidity: Watson wryly suggested to Lashley that he endow two professorships in sexology "with a man professor to teach young men how to slow down and a woman professor to teach gals how to speed up and to have an orgasm."[18] When it came to the "primitive" tribes of the third world, however, the problem that Watson and Lashley discussed was how

---

[11] Lashley to Watson, Oct. 18, 1956, Lashley Papers, Box 2, folder 3, UFG. These "shots of testosterone" were apparently part of the treatment for the ailments Lashley suffered late in his life, which included an extreme form of anemia and weakening of the spinal vertebrae.

[12] Lashley to Watson, Aug. 7, 1953, Lashley Papers, Box 2, folder 3, UFG.

[13] Lashley to Watson, June 16, 1954, Lashley Papers, Box 2, folder 3, UFG.

[14] Lashley to Watson, May 1, 1953, Lashley Papers, Box 2, folder 3, UFG.

[15] Watson to Lashley, Sept. 16, 1953, Lashley Papers, Box 2, folder 3, UFG.

[16] Watson to Lashley, Oct. 3, 1955, Lashley Papers, Box 2, folder 3, UFG.

[17] Lashley to Watson, Oct. 27, 1955, Lashley Papers, Box 2, folder 3, UFG.

[18] Watson to Lashley, Sept. 16, 1953, Lashley Papers, Box 2, folder 3, UFG.

the tribal elders managed to rein in their women's rampant sexuality. The issue was not frigidity, but rather how to make the women frigid. Watson and Lashley imagined they could exploit such sexual resources: "Why don't we go to China," Watson asked,

> and buy a few gals and give them a couple of dollars when we leave. I am simple minded enough to want a woman around but not always the same woman.[19]

A recurrent topic of conversation in the correspondence was female circumcision, the procedure which was supposed to hold in check the problem of "primitive" female sexuality. "I read lately that some of the primitive tribes circumcised the females as well as the males," Watson told Lashley. "When I looked it up with the few books I have – it said often removed the clitoris too. Do you mean to tell me that even primitives didn't know the goose that lays the golden egg?"[20]

Lashley responded that he had read that "cutting off the clitoris is practiced by a number of West African tribes," as part of the initiation ceremony at puberty. In the women of a certain East African tribe, he continued,

> the labia minor are developed as flaps which hang almost to the knees. It is not impossible that there are racial differences in innervation likewise. On the other hand, primitive peoples sometimes have little regard for their women except as laborers.[21]

A letter from Lashley a year later gave Watson a fuller description of female circumcision. Lashley quoted a passage from an account critical of the custom written by a chief of the southern African Basuto tribe, which practiced it. The practice, Lashley reported, involved not removing the clitoris, but covering it with a flap of skin; the Basuto chief noted that " '[t]he performance of this rite tends to encourage chastity among the women, for a circumcised girl can know little of the joys and passions of physical love.' " Lashley's quotation of the account continued,

> "It perhaps can be said that the circumcision of women not only denies the girl great pleasure and joy in the sexual act, but must in consequence lessen the happiness and exaltation of the man, and thus shut out any upliftment of the spirit – lying with a woman then becomes a selfish, rather than a mutual pleasure."

The Basuto chief noted that this custom was the cause of much male homosexuality, " 'which is brought to flower by the living conditions imposed upon African mine workers by the white man.' " Lashley remarked that the "practice is wide-

---

[19] Watson to Lashley, July 7, 1954, Lashley Papers, Box 2, folder 3, UFG.
[20] Watson to Lashley, n.d. (probably May 1953) Lashley Papers, Box 2, folder 3, UFG.
[21] Lashley to Watson, Sept. 18, 1953, Lashley Papers, Box 2, folder 3, UFG.

spread through Africa, from the Berbers of the north coast, the MauMau, that are giving the British so much trouble now in the East, to the Basuto in the South."[22]

Nearly nine months after this letter, Lashley provided Watson with further information on female circumcision, this time from a medical worker stationed in southern Africa who had taken photographs of the procedure. "In this case it involved cutting out a large wedge of flesh, including the clitoris," Lashley wrote. "The explanation of the man who took the pictures is that the operation is intended to make the woman frigid and so less likely to go with other men than the husband."[23]

To judge from their sustained interest in female circumcision over several years, I think it is fair to say that Lashley and Watson were fascinated by reports on the practice. Lashley's recounting of his investigations into the matter, and Watson's always earthy responses to them, illustrate the worldview in which they were operating: women's sexuality could not be considered apart from race. As scientists and as men, they confronted "the problem of frigidity," but it was always the frigidity of white women they were considering. White women were possessed of a sexuality needing enhancement; and the problem was how to make these women, "their own" women, more receptive to their advances.

However, if what a man really wanted was sex, then nonwhite, non-Western women were clearly the best sources. This belief was echoed and reinforced by the reports on circumcision, which suggested that there were women whose sexuality was so powerful that it threatened to overflow its boundaries and needed to be controlled. In Africa and China, the problem was not frigidity; in fact it was just the opposite. The leaders of the "primitive" tribes" were concerned with how to make their women frigid. Thus race and racial difference became an integral element both of Lashley's semiscientific reports of female circumcision and of Watson's obsessive ruminations on women's sexuality.

The racial theme emerged even more clearly in the second part of the correspondence. In May 1955, Lashley announced his intention to make a visit to Jamaica; the day after his return, seemingly unable to wait to tell Watson about what he had found, Lashley wrote:

Yesterday I got back from Jamaica after a month at sea and in the islands. Jamaica is the tropical paradise for which I was searching: wonderful climate and scenery, no insects, cleaner than any American city I know, cheap, almost slave labor (3 servants for a total of $6 per week) and even dusky maidens (I had to throw one out of my room my first night there).

He continued:

[22] Lashley to Watson, Sept. 6, 1954, Lashley Papers, Box 2, folder 3, UFG.
[23] Lashley to Watson, May 29, 1955, Lashley Papers, Box 2, folder 3, UFG.

Nevertheless I shall not retire there. It was my first experience in an all black country with no color line. I could not quite stomach dining in the best hotel as the only Caucasian with 26 negroes.[24]

This was the same experience that he related in letters to Taliaferro, Klüver, and Bishop. That he told the same story to different correspondents shows that it had touched a deeply sensitive nerve. Even a year after his trip, Lashley remained shaken by what he had seen in Jamaica. He told Watson, "I really don't understand how you can endure the northern winters. North Florida is bad enough and if I can ever find a tropic isle that is not lousy with niggers, I shall take root there."[25]

But Lashley's feelings about race relations were not limited to his experiences in foreign countries. He was also committed, as we have seen from his 1950 interview with Anne Roe, to the idea of a "color line" in his own country. To Watson he wrote in December 1956, "Made a flattering discovery in New York. There are 10 or a dozen Lashleys listed in the N.Y. telephone directory. They are practically all negroes, which probably means that some of my ancestors were extensive slave owners."[26] Lashley was so impressed by his find that, as he had done with his Jamaica excursion, he told the same story to others besides Watson. To cousins who owned a sugar plantation in the Dominican Republic, Lashley wrote:

The New York telephone directory lists a number of Lashleys. I discovered when I was there in Dec. that they are practically all black. Some of my ancestors were undoubtedly slave owners and I am quite set up about it.[27]

Lashley was obviously pleased by this discovery about his forebears. A month later he wrote to Watson,

Sorry to hear that you have had a bad winter and so much cold but perhaps it serves you right for abandoning your native south and staying among the Yankees. Me, I am getting to be more of a rebel every day, ready to go to any lengths to stop integration.[28]

The last decade of his life Lashley spent in the Jim Crow South, trying to make sure that things remained the same as they had always been. By the 1950s, the racial hierarchy of early twentieth-century America was beginning to be dismantled by the civil rights movement. The decision in *Brown v. Board of Education,* desegregating the public schools, came in 1954; the Montgomery bus boycott in 1955. His letters from the 1950s show that for Lashley these political and social developments were more than a Zeitgeist or a vague cultural background;

[24] Lashley to Watson, Oct. 27, 1955. Lashley Papers, Box 2, folder 3, UFG.
[25] Lashley to Watson, Oct. 18, 1956, Lashley Papers, Box 2, folder 3, UFG.
[26] Lashley to Watson, Dec. 26, 1956, Lashley Papers, Box 2, folder 3, UFG.
[27] Lashley to Rhoda and Jan Jongsma, April 9, 1957, Box 1, folder 12, UFG.
[28] Lashley to Watson, May 11, 1957, Lashley Papers, Box 2, folder 3, UFG.

they illustrate how heavily the threat of integration weighed upon his mind, how consciously and consistently he worried about it. Moreover, the letters show that Watson, in his own way, echoed and affirmed Lashley's opinions; his remarks about women's sexuality depended on his holding the belief in racial hierarchy and racial difference that Lashley made explicit.

Lashley's backward-looking political point of view was matched by a scientific conservatism that he also shared with Watson. Both lamented the current wretched state of psychological science and mourned the loss of an earlier era, before certain destructive influences had taken hold. Lashley believed that training for clinical practice was taking over the field; despite the growing membership of the American Psychological Association, he told Watson, there had been more "good research men" in 1918.[29] He and Watson commiserated over the resurgence of interest in the "soul" that they saw in their respected colleagues:

> There is . . . a distressing revival of outmoded psychology among neurologists. Sir Fredric Walshe, the most influential clinical neurologist in England, has come out for the SOUL, as did Sherrington in his senile days. Eccles, one of the best neurophysiologists, is nearly as bad and Penfield, the only neurosurgeon who is doing decent research, seems to think that there is a little man in the reticular formation.[30]

Watson sympathized: "Sometimes when you look back over your life you wonder if anything is worthwhile. How could mysticism have gotten such a hold. I am amazed at the neurologists."[31] Watson told his friend that he had even stopped taking the covers off the new psychological journals anymore; and Lashley concurred:

> Just got the program of the APA meeting in Chicago next September. Christ, what a mess. There seem to be fewer "Presidential Addresses" this time. Last year I believe there were 15. I note there is now an "American Catholic Psychological Association." In the 30 pages of titles I cannot find one paper that I would listen to.[32]

As they nostalgically remembered an earlier, better day in science, Lashley and Watson also wanted to turn back the clock on society. Lashley himself suggested that his own social views were so conservative, or rather, retrogressive, that they could not be located on the spectrum of American politics. Hence his unwillingness to vote in the 1956 elections: "I don't like Ike and Nixon stinks of McCarthy," he told Watson in November 1956, "but the alternatives were worse. . . . The two party system is more efficient than the French method but, since I agree

[29] Lashley to Watson, May 1, 1953, Lashley Papers, Box 2, folder 3, UFG.
[30] Lashley to Watson, October 27, 1955, Lashley Papers, Box 2, folder 3, UFG.
[31] Watson to Lashley, n.d., probably 12/56, Lashley Papers, Box 2 folder 3, UFG.
[32] Lashley to Watson, July 20, 1956, Lashley papers, Box 2, folder 3, UFG.

with neither party, for me the American is no different from the Russian one party system."[33]

Lashley's frequent protestations that he was alienated from society and that his scientific work contributed nothing toward human betterment have been interpreted as statements of his scientific neutrality. His pronouncements that his model of brain function lacked practical applicability have been understood to mean that Lashley had no political beliefs and no hopes for society, or at least none that polluted his scientific purity. In fact, however, his professed societal alienation simply meant that he was opposed to changing society. The ideal society for Lashley was the one he was living in, with the "Negroes" in their place, and the whites in theirs.

He expressed this ideal clearly in one letter to Watson. After Watson facetiously suggested that Lashley use his extra income for endowing professorships in sexology, Lashley answered half jokingly,

> I am shocked to find you in the spirit of a reformer! The professorships that you suggest would be just another form of uplift – to greatly improve the human lot and probably increase the breeding rate.

Then Lashley became more serious:

> Myself am a believer in *things as they are.* As I look back over the more than 40 years that I have spent in research, I am glad to see that there is not one contribution of practical value. And giving money for research in brain functions almost assures that it cannot be used for any useful purpose.[34] [emphasis in original]

For Lashley, then, to be neutral was implicitly to oppose social change, to counter integration, to keep the hierarchy in place. When this detachment from society was combined with the hereditarianism that flowed from Lashley's scientific work, his powerful argument for the status quo falls into place. Society should not be changed, segregation should not be challenged, because the social order really could not be otherwise. Biologically determined patterns of behavior had made it the way it was. Social change was literally an impossibility.

Lashley's Last Year

In May of 1957, Lashley abruptly announced to Watson:

> I hope that your blood pressure is down for here is a shocker. I am getting married in June to a charming widow who, by all available evidence, is at least 25 years my junior. . . . She is Hungarian, Ph.D Budapest, niece of

[33] Lashley to Watson, Nov. 12, 1956, Lashley Papers, Box 2, folder 3, UFG.
[34] Lashley to Watson, Sept. 18, 1953, Lashley Papers, Box 2, folder 3, UFG.

former Prime Minister, Imradi, and widow of Paul von Schiller, whom I brought to the Labs as research psychologist in '47. She reminds me in many ways of Rosalie [Rayner], on whom I can now confess I was developing a crush before you carried her off.[35]

Lashley's second wife was Claire Imradi (or Imredy) Schiller (1912–1988); the two had become acquainted in the early 1950s when they worked together to finish and publish the book Paul Schiller had begun, *Instinctive Behavior: The Development of a Modern Concept*. This book, which comprised an English translation of the work of European ethologists, was published by International Universities Press in 1957, translated and edited by Claire Schiller and with an introduction by Karl Lashley.[36] Claire Schiller was a highly accomplished woman in her own right, as Lashley's letter suggests: in addition to her doctorate from the Royal Hungarian Petrus Pazmany University, she received a diploma in English language and literature from King's College of the University of London. In 1949, following her first husband's death in a skiing accident, she became a teacher at the Bartram School in Jacksonville to support herself and her young daughter Christina (b. 1942).[37]

Watson seemed delighted to hear this news of his friend's upcoming nuptials. "Most sensible thing I ever heard of your doing. Congratulations. I wish you both a lot of happiness. May you flourish like a green bay tree. Be fruitful and multiply!"[38]

Indeed the last years of Lashley's life seem to have been among his happiest. Though retired from the Yerkes Laboratories directorship in 1955, he continued to write, to lecture, and to stop in at the labs, as well as to sail his boat and play his cello. The summer of 1957, following their wedding, he and Claire took a driving trip around the United States, revisiting scenes of Lashley's boyhood in Seattle and Alaska. This they did despite Lashley's frail health, a constant subject in his letters to Watson: Lashley suffered from extreme anemia and weakening of the spinal vertebrae.[39] Nevertheless, he wrote to Watson from his honeymoon trip:

[35] Lashley to Watson, May 11, 1957, Lashley Papers, Box 2, folder 3, UFG.

[36] Claire H. Schiller, trans. and ed., *Instinctive Behavior: The Development of a Modern Concept* (New York: International Universities Press, 1957). Contributions by D. J. Kuenen, Konrad Lorenz, Nicholas Tinbergen, Paul H. Schiller, Jacob von Uexküll; introduction by Karl S. Lashley.

[37] On Claire and Paul Schiller, see Donald A. Dewsbury, "Paul Harkai Schiller," *The Psychological Record* 44 (1994): 307–350.

[38] Watson to Lashley, May 12, 1957, Lashley Papers, Box 2, folder 3, UFG.

[39] In February 1954, as he arrived at Harvard from Florida in order to do some teaching, Lashley collapsed and was hospitalized. As he later wrote to Watson, "I had hoped to see you again when I went North last winter but landed in the hospital as soon as I got to Boston. Spent February in the Mass. Memorial and March here [Jacksonville] in St. Vincent's. One of the damned mysterious anemias; guesses ranged from T.B. to cancer of the pancreas – all wrong, so I keep going on three dollars worth of cortisone per day. Think how much good liquor that would provide!" (June 16, 1954, Lashley Papers, Box 2, folder 3, UFG). The diagnosis was officially "acquired hemolytic anemia," to be treated as first with cortisone, then by splenectomy, which Lashley underwent in

You were right, the most sensible thing I ever did. Has taken 20 years off my age. We have driven over 4,000 miles – I climb mountains and try to swim again – for the first time since Bird Key. Karl Lashley. (Spilled my gin here.)[40]

Lashley and Watson met again in New York in September 1957, and later that autumn Lashley wrote to his friend that his health was much better, "though I am never completely free of pain."[41] Their last few exchanges focus chiefly on their usual concern with the intellectual poverty of psychoanalysis, though much of the male bonding and posturing evident in the earlier letters are absent. Lashley, however, kept up his habitual attitude of self-deprecation, ending his last letter to Watson:

A group of my former students is getting out a volume of my collected – or rather, selected – papers. It is supposed to be a surprise for me but I have been consulted on everything but what is to be included. I fear that its chief effect will be to show that I have changed my opinions every two years. Well – God preserve me from all reverence for authority, even Thine and mine own – should be the scientist's only prayer.[42]

The second summer of their marriage, Karl and Claire took Christina to Europe, to visit Claire's parents in Portugal and to tour Italy, France, and England by car. They sailed from New York on June 18, 1958. In early August, while they were in Poitiers, France, Lashley succumbed to his chronic ailments; he died on August 7, at the age of 68. He and Claire had been married for fourteen months.

On September 26 of the same year, Ruth Lieb, Watson's secretary, sent a telegram to Claire Lashley:

JOHN WATSON DIED PEACEFULLY THIS AFTERNOON FUNERAL SATURDAY MORNING AT ELEVEN O'CLOCK NORFIELD CHURCH WESTON CONN RUTH[43]

1955. The cortisone treatments apparently softened his spinal vertebrae, and he suffered spinal dislocations over the next several years. (See Frank A. Beach, "Karl S. Lashley, June 7, 1890– August 7, 1958," *Biographical Memoirs of the National Academy of Sciences* 35 [1961]: 194–95.) In 1956, Lashley wrote Watson: "Last May I got a bit too spry – went for a ride in my neighbor's speed boat and busted my spine again when he hit some rough water. Have had a painful time since but manage to keep it under control with aspirin and gin. My vertebrae seem to be pretty badly moth-eaten." (July 20, 1956, Lashley Papers, Box 2, folder 3, UFG.) In addition to the cortisone and the home remedies Lashley mentions, the treatment for his ills included blood transfusions and shots of testosterone.

[40] Lashley to Watson, July 6, 1957, quoted in Donald A. Dewsbury, "The Watson-Lashley Correspondence of the 1950s" *Psychological Reports* 72 (1993): 266–7.

[41] Lashley to Watson, November 21, 1957, Lashley Papers, Box 2, folder 3, UFG.

[42] Lashley to Watson, February 4, 1958, Lashley Papers, Box 2, folder 3, UFG.

[43] Telegram quoted in Dewsbury, "Watson-Lashley Correspondence," 268.

In a letter following this telegram, Ruth Lieb wrote to Claire Lashley, "words seem inadequate to comfort one – I wish I knew some magic ones to help you over the emptiness you must be feeling now."[44]

Widowed for a second time, a grief-stricken Claire Lashley wrote to Heinrich Klüver:

> One moment, it seems all right almost – that after a happy year, the happiest of his life, he should be allowed his rest, his peace, knowing at last what he wanted all his life so terribly to know, having provided for me and Christina – and the next moment it all goes to pieces, like a house of cards, a fabrication, leaving only boundless unhappiness.[45]

To the list of the ironies of Lashley's life with which I opened Chapter 1, we may then add one more: an unquestionably brilliant, pioneering brain scientist, widely admired for his work, whose neutral facade concealed private political views that were less than admirable. Let me emphasize once again that my purpose here is not to condemn Lashley for his private beliefs, nor to hold him to a late twentieth-century standard. It is, rather, to try to unravel the mass of contradictions that comprised Lashley's life, to see that life in its complexity, feel the force of its ironies, and make sense of it. Lashley's tragedy was that he belonged to a social order that was fast disappearing, to a world that was slipping away before his eyes; his life's desire was to hold onto that world as long as possible, to grasp its comforts and its familiarity, even as the ground was shifting underneath him. He always showed, as he suggested in his last letter to Watson, a deep irreverence for any kind of authority; Lashley's iconoclasm became his trademark; and yet his allegiance to the old order, to the way of life he had always known, was absolute. His theory of the brain, supported by the twin pillars of his neutral stance and his hereditarian conception of intelligence, comprised his argument against social change.

That Lashley's political beliefs remained firmly rooted in the past, even as he pushed forward the frontiers of brain research, is the most tragic of his ironies. But as much as his surgical methods and laboratory practices, his longing for the past was a part of his neuropsychology, and helped to shape its rhetoric and substance. Only by understanding that longing, by reading Lashley's personal letters along with his published papers, can we make sense of the meaning and significance of his scientific contribution.

---

[44] Lieb to C. S. Lashley, October 19, 1958, quoted in Dewsbury, "Watson-Lashley Correspondence," 268.

[45] Claire Lashley to Heinrich Klüver, October 30, 1958, "Klüver" folder, Claire Schiller Lashley Papers, UFG. [The original is wrongly dated.]

11

Genetics, Race Biology, and Depoliticization

Prologue

While Lashley established his political point of view early in his career, and was never able or willing to shake it, the significance of maintaining such beliefs did change radically during the first half of the century. In the first few decades, through the 1920s, racist beliefs, even virulently racist, would hardly have been surprising. Racism was the expected and constant concomitant of the study of human biology, psychology, and anthropology. By the 1930s, however, this accepted norm had begun to erode; and by the early 1940s, for a variety of reasons, an atmosphere of antiracism had arisen in the biological and social sciences.[1] It had become distinctly unfashionable to express racist views publicly or in one's science; life scientists were trying furiously to distance themselves from the notion that certain races were inferior to others.

In this chapter I will argue that even amid this shift toward the apolitical, the hereditarian assumptions that historically had always marked scientific studies of human differences persisted even after 1940. Lashley's work, particularly his participation in the NRC Committee on Human Heredity and his late interest in behavior genetics, conforms to this broader trend: a surface of neutrality, undergirded by an unshaken belief in the power of heredity.

Race Biology and Its Depoliticization

One of the classic examples of early twentieth-century scientific racism is the work of the anatomist and breeder Charles R. Stockard (1879–1939). Professor of anatomy at Cornell University from 1911, Stockard was the director of the so-

---

[1] See Franz Samelson, "From 'Race Psychology' to 'Studies in Prejudice': Some Observations on the Thematic Reversal in Social Psychology," *Journal of the History of the Behavioral Sciences* 14 (1978): 265–278; George W. Stocking, Jr., *Race, Culture and Evolution: Essays in the History of Anthropology* (New York: Free Press, 1986); Carl N. Degler, *In Search of Human Nature: The Decline and Reversal of Darwinism in American Social Thought* (New York: Oxford University Press, 1991); and Elazar Barkan, *The Retreat of Scientific Racism: Changing Concepts of Race in Britain and the United States Between the World Wars* (New York: Cambridge University Press, 1992).

called Experimental Morphology Farm at the Cornell Medical School from 1926 until his death. There he carried out breeding experiments on dogs, intended to correlate hereditary anatomical features, hormonal imbalances, and behavior. His work in comparative anatomy and physiology was well respected and influential in its day; Lashley, for example, cited it more than once in his own scientific writings. But Stockard also meant to apply his findings directly to the human species; he believed he had found compelling scientific evidence that race mixing should be prevented.

In 1931, Stockard published *The Physical Basis of Personality,* showing, he argued, that alteration in glandular function could produce distortions in the growth and form of dogs. He proved the same was true for human beings by demonstrating the striking similarities in facial appearance of dogs and people who suffered similar glandular ailments. "[I]t is of great significance," Stockard proclaimed in the preface to this work, "that certain human freaks practically parallel in their growth and form these diversified canine types."[2] Photographs of the faces of the human "freaks," each matched to a picture of its canine counter-part, drove the point home.

Stockard was interested not only in individual differences, but in differences between groups as well. He speculated that different races of man, like different breeds of dog, possessed different "racial personalities" and that different nations had different "nationalistic personalities." (These, however, occurred only among the "white and yellow races" – which were, presumably, the only ones advanced enough to have nations.)[3] He theorized that these different group personalities could be traced to shared differences in brain anatomy: "We may some day be surprised to find that, within normal limits, brain shape is more significant for mental quality and ability than brain size."[4]

Ten years later, in 1941, Stockard's *The Genetic and Endocrinic Basis for Differences in Form and Behavior, As Elucidated by Studies of Contrasted Pure-Line Breeds and Their Hybrids* was published posthumously by his colleagues at the Experimental Morphology Farm.[5] Lashley cited this work in his 1949 article "Persistent Problems in the Evolution of Mind."[6] Far from reversing any of his earlier conclusions, Stockard here extended his arguments. Not only did the interactions of genes and hormones produce racial differences in personality in dogs and human beings, but these differences provided ample justification for the prevention of "race crossings." In the most remarkable passage of the book, Stockard wrote that while race crossings in dogs have been guided by "a master hand,"

[2] Stockard, *The Physical Basis of Personality* (New York: W.W. Norton, 1931), vii.
[3] Ibid., 302.
[4] Ibid., 279.
[5] American Anatomical Memoirs 19 (Philadelphia: Wistar Institute, Oct. 1941).
[6] Lashley, "Persistent Problems in the Evolution of Mind," *Neuropsychology of Lashley,* 477.

[n]o such force regulates the mongrel mixing of human beings. . . . On the contrary, individuals carrying different degrees of those tendencies [toward different body shapes] are constantly being absorbed into the general human stock, possibly to render the hybridized races less stable and less harmonious in their structural and functional complexes than were the original races from which they were derived. Mongrelization among widely different human stocks has very probably caused the degradation and even elimination of certain human groups; the extinction of several ancient stocks has apparently followed very closely the extensive absorption of alien slaves.

Stockard continued:

If one considers the histories of some of the south European and Asia Minor countries from a strictly biological and genetic point of view, a very definite correlation between the amalgamation of the whites and the negroid slaves and the loss of intellectual and social power in the population will be found. Contrary to much biological evidence on the effects of hybridization, racially prejudiced persons, among them several anthropologists, deny the probability of such results from race hybridization in man.[7]

"There is no doubt," Stockard concluded by saying, "that unstable individuals with structural and functional disharmonies arise from crosses between contrasted breeds of dogs," and the same disharmonies are evident in the hybridized offspring of other species, including human beings. Stockard's study was, of course, completely objective:

Certainly no answer to these debatable and very important questions can be scientifically arrived at by any method other than careful experimentation on higher mammals, and, in the light of such experiments, an impartial study of the unregulated human results.[8]

But by the early 1940s Stockard's virulently racist views were falling into disfavor, at least in public scientific circles. Lashley's attempt to create a politically neutral science was mirrored by scientists trying to depoliticize their studies of race. Many biologists and social scientists began to moderate their allegiance to the idea of hereditary mental differences between the races. This they did not by denying that there were races or differences between them, but by denying that these differences formed a barrier to inbreeding.

There were a number of strategies that scientists adopted to depoliticize their study of race. The first was to focus entirely on hereditary physical difference, which would engender, they believed, little heated political dispute. Second, they took an environmentalist approach, arguing that there were mental differences

[7] Stockard, *Genetic and Endocrinic Basis,* 37–8.
[8] Ibid.

between the races but that they were due entirely to cultural circumstance, and so could easily be modified. The third was to profess agnosticism about racial differences, claiming that nothing scientific was as yet known about them, and that more research was necessary. Closely related to this last strategy was the belief that the only truly scientific approach to the question of difference was to concentrate on individuals, and to abandon the search for group differences. These three approaches are exhibited in *Scientific Aspects of the Race Problem* (1941), a volume of essays by biologists, anthropologists, and psychologists.[9]

An example of the first approach is provided by one of the more conservative contributors to the volume, the geneticist H. S. Jennings, Lashley's teacher. Citing Stockard and his own *Biological Basis of Human Nature* (1930), Jennings wrote that certain races "may yield combinations that are not harmonious."

> Large organs from one parent may be combined with small ones from the other, or there may be other disharmonious combinations. . . . It is clear that disharmonious combinations do occur in human beings; such may well be the results of crosses between very different individuals . . .[10]

Although he believed in disharmonious racial combinations, Jennings made it clear that he was talking only in terms of physical traits. Some races were not meant to have offspring together because they were just too disparate in physical size, not because their psychologies were incompatible. Jennings, in fact, professed agnosticism when it came to mentality; his essay dealt only with "the material side" of human genetics and heredity. He depoliticized his argument by not discussing mental traits or "race psychologies" at all, but by concentrating solely on body size and shape.

The second method of depoliticizing the study of race, the environmentalist approach, was used by only one contributor to the volume, the anthropologist Ales Hrdlicka. He began by stating that race was a purely physical classification: "No attempt has ever been made, nor would it be possible, to distinguish and define 'race' on the basis of either physiological or mental abilities."[11] In fact, however, this is precisely what Hrdlicka goes on to do in the rest of his essay. There were indeed differences between the races, but these differences were entirely culturally and environmentally caused. The power of the environment to shape brain and behavior was such that both physical and mental differences between the

---

[9] H. S. Jennings, Charles A. Berger, Dom Thomas Verner Moore, Ales Hrdlicka, Robert H. Lowie, and Otto Klineberg, *Scientific Aspects of the Race Problem* (Washington, D.C.: The Catholic University of America Press, 1941). Preface by Bishop Joseph W. Corrigan, Rector of the Catholic University of America.

[10] H. S. Jennings, "The Laws of Heredity and Our Present Knowledge of Human Genetics on the Material Side," *Scientific Aspects of the Race Problem* (Washington, D.C.: Catholic University of America Press, 1941), 71–2. See also Jennings, *The Biological Basis of Human Nature* (New York: W. W. Norton and Co., Inc., 1930); on Jennings and race, see Barkan, 189–210.

[11] Ales Hrdlicka, "The Races of Man," *Scientific Aspects*, 161.

races must exist. To pretend that these differences did not exist "would be to paralyze all the tenets of the biology of human plasticity, inheritance, evolution, eugenics and turn to stagnation."[12] But equally because of the power of the environment, these racial differences could be eradicated or altered: the more "belated" races (Hrdlicka preferred not to say "inferior" or "superior") could with some effort "catch up" to the more "advanced" races "under proper sustained stimuli and above average exertions."[13]

Hrdlicka believed that the trend toward race mixing was already working toward the erasure of existing differences and the creation of new races:

> Differentiation, mixture, blending, integration, raciation, are potent and unceasing processes, and as Man is still plastic they will have their results. In twenty–thirty thousand years, in consequence, there will be racially, as well as otherwise, a new human world, with bare traces or remnants of the present.[14]

Hrdlicka's environmentalist answer was also given by L. C. Dunn and Theodosius Dobzhansky in their 1946 book *Heredity, Race and Society*.[15] There they argued against the deleterious effects of race crossing, writing that "[t]here is no doubt that civilization leads slowly but inexorably to the breakdown of race divisions."[16] This was a trend that should not, and could not, be resisted. In Dunn and Dobzhansky we see the almost complete opposite of Stockard's and Lashley's beliefs. While Stockard believed that race "mongrelization" led to "disharmonious" offspring, and Lashley tried to keep the different races apart, Dunn and Dobzhansky argued that race mixing could lead to improvements in the next generation:

> Contrary to the opinion vociferously expressed by some sincere but misguided people, such a trend is not biologically dangerous. Mixing of closely related races may even lead to increased vigor.[17]

When it came to races at the extremes of the spectrum, however, Dunn and Dobzhansky were more cautious in their assessment:

> As for the more distantly separated races, there is no basis in fact to think that either biological stimulation or deterioration follows crossing. The widespread belief that human race hybrids are inferior to both of their parents and somehow constitutionally unbalanced must be counted among the superstitions.[18]

[12] Ibid., 169.
[13] Ibid., 170.
[14] Ibid., 186.
[15] Dunn and Dobzhansky, *Heredity, Race and Society* (New York: The New American Library, 1946).
[16] Ibid., 129.
[17] Ibid., 130–1.
[18] Ibid.

Whether or not they fully supported the idea of mixing racial "extremes," Dunn and Dobzhansky were still opposed to Stockard's basic assumption: that mental abilities, as much as physical traits, were hereditarily determined. For Dunn and Dobzhansky, the most important feature of human evolution was that "the human species as a whole has developed away from genetic specialization and fixity of behavior, and toward educability."[19] Since Dunn and Dobzhansky believed that environment determined behavior and mentality more strongly than did genetics, the races could be mixed more or less indiscriminately. "This does not mean," Dunn and Dobzhansky concluded, "that the genetic differences among men do not affect their mentality. But from the vantage point of evolutionary biology we can see that such differences are not fundamental. Far more important is the fact that human capacities are developed by training from childhood on."[20]

In direct opposition to Stockard's notion that different races had different genetic "personalities," Dunn and Dobzhansky believed that "cultural differences" bore much greater responsibility for these "so-called 'race psychologies'" than did biological heredity.[21] Despite the seeming liberality of their views (especially in contrast to Stockard and Lashley), Dunn and Dobzhansky did not deny that racial differences existed. In fact they did believe that there were racial differences in personality, although they thought that these were culturally, not biologically caused.

Aside from the materialist and the environmentalist solutions, other contributors to *Scientific Aspects of the Race Problem* took the third route toward depoliticization – an agnosticism about the existence of race differences in intellectual capacity. They argued that science must focus on the differences between individuals, not between races. In his article about race differences in intellect and morality, Robert H. Lowie wrote that "[i]n neither case can observed differences be interpreted in racial terms."[22] Otto Klineberg, in his contribution on the mental testing of racial groups, concluded that

> [i]t appears certain that both heredity and environment enter into the determination of individual differences in the test scores, but the influence of heredity upon group differences has so far not been demonstrated.[23]

The key words here are "so far." These writers were not by any means denying that racial differences in intellect could exist. Physical differences between races were clearly hereditary, and clearly both physical and mental differences between individuals were at least in part genetically determined. Mental differences between groups, however, they preferred to leave an open question, still subject to

---

[19] Ibid., 133.

[20] Ibid.

[21] Ibid., 134.

[22] Robert H. Lowie, "Intellectual and Cultural Achievements of Human Races," *Scientific Aspects*, 244.

[23] Otto Klineberg, "Mental Testing of Racial and National Groups," *Scientific Aspects*, 284. Klineberg was strongly influenced by the liberal anthropologist Franz Boas.

scientific discussion. More evidence was needed to decide the answer. Lowie, for example, wrote that

> [t]he suggestion that racial strains may differ in variability as to particular aptitudes, temperament, etc., cannot be brushed aside as worthless, but it should be treated as what it is, – not as a scientifically authenticated conclusion, but as a very humble initial hypothesis in a field in which literally nothing is *known*.[24] [emphasis in original]

The call, then, was not to shun the whole area of race differences, but to conduct further – and this time, authentically scientific – investigations into it.

The contributors to this volume saw that the integration of the races was a problem; and it was a problem to be treated and solved by science. If there was any shift in thinking about the "problem of the color line" during the first half of this century, it was a shift toward a self-consciously apolitical, moderate, scientific discourse. The agnostic's answer to the race problem relied on the findings of science for a solution, and those scientists who addressed this issue made it absolutely clear at the outset that they, in contrast to their predecessors, were truly "scientific."

Pure Science and Hereditarianism

By the 1940s, then, most anthropologists and psychologists, and a few geneticists, were able to cast off the idea of hereditary mental differences between groups, and thereby depoliticize their science of race. But for many other geneticists, whose science had rested squarely on the hereditarian ideal since the turn of the century, the story was different and more complicated. Since by the last decade of his life Lashley thought that the answers to the puzzle of brain and behavior lay in the genes, and since it was to geneticists, rather than to psychologists, that he showed the greatest affinity, I turn now to geneticists' response to the depoliticization issue, and to Lashley's participation in that response.

The story hinges on two episodes: first, Lashley's decade-long membership in the NRC Committee on Human Heredity; and second, John Paul Scott's development of behavior genetics, to which Lashley's work was a much more successful parallel. Both episodes show that, like their counterparts in psychology and anthropology, geneticists were trying to construct a "pure" and politically neutral science of human difference, without the traditional overlay of eugenic and social-betterment rhetoric. But even in this new genetics, as we shall see, the older hereditarian ideals persisted. Like Lashley, his colleagues on the Heredity Committee and the geneticist Scott worked to identify the innate basis of behavior in the name of pure science.

Lashley joined the NRC Committee on Human Heredity in 1938, after the

[24] Lowie, 226.

resignation of the psychologist Robert S. Woodworth, and remained a member until the committee was disbanded in 1947. The committee's work in the years during and after the Second World War clearly reflects Lashley's interest in developing a "pure" and hereditarian approach to psychology and biology. Members of the committee sought to achieve this goal by drawing a sharp distinction between their own pure genetics and traditional, socially involved eugenics. The historian Diane Paul, among others, has argued that while outright eugenist rhetoric may have faded during this period, the hereditarian thinking of the eugenists remained, absorbed into the more technical disciplines of behavioral and medical genetics. The eugenists, for example, had been concerned with the inheritance of mental traits like feeblemindedness, believing that improvement of physical form would follow from enhancement of mental power. In the same way, later medical geneticists were interested in the genetics of intelligence.[25]

The work of the Committee on Human Heredity exemplifies this explicit shift away from eugenics, but an implicit maintenance of all its hereditarian assumptions. A prospectus of the committee said that its purpose was to base human genetics on facts, since genetics was basic to human welfare. But unlike eugenics, human genetics should be purely objective, divorced from "social implications" except as they followed immediately. The committee nonetheless maintained the eugenists' interest in intelligence, addressing problems of inheritance of the "so-called 'mental' traits in man."[26] Moreover, the original membership of the committee seriously belies its detachment from eugenics: in 1939, in addition to Lashley, the committee included the prominent eugenists Charles B. Davenport, Harry H. Laughlin, and Paul Popenoe. The following year the committee was reorganized, and these three were excluded. But the grants that the committee considered for support still had a strong eugenist flavor; the committee must have been seen by applicants as receptive to such projects.[27]

In 1940, for example, the committee considered grant applications from one J. H. Landman of the City College of New York, "a eugenist, having made somewhat of a reputation with my work on subnormal people and sterilization," for a study of "human genius." An application also came in from Abraham Myerson of Boston State Hospital, the chairman of the Committee on Heredity and Eugenics of the American Neurological Association. He proposed a study of psychosis in "distinguished," "ordinary" and "criminal" families of New England. In his application he wrote that psychosis "peppers the entire human field," but that this "of course, is not at all an argument against sterilization."[28]

Moreover, the committee continued to support Davenport even after he was no

[25] See Diane Paul, "The Rockefeller Foundation and the Origins of Behavior Genetics," *The Expansion of American Biology*, 262–283.

[26] "Prospectus of Committee on Human Heredity," Lashley Papers, Folder "Committee on Human Heredity of the NRC," Yerkes.

[27] Ibid.

[28] Folder "Applications for Grants from the Committee on Human Heredity," Lashley Papers, Yerkes.

longer a member. In 1940, Davenport applied for a grant to study the parts of the body that were under "gene control." Apparently this grant was at first denied, but after both Lashley and Sewall Wright wrote letters to the chairman asking for a reconsideration, Davenport received his grant.[29] In a 1941 letter to the chairman of the committee, a young Columbia zoologist named James Neel wrote that he could best serve his country by studying human genetics "along eugenical lines."[30] (Neel later became a medical geneticist at the Institute of Human Biology of the University of Michigan.)

The work of the committee, then, demonstrates the continuity between eugenics and the "purer" sciences of behavioral and medical genetics. Its members maintained the eugenists' interest in the inheritance of psychological characteristics like intelligence. Nonetheless, it is clear that the committee was supposed to be putting genetics on a more "scientific" footing; and its concern with pure science is especially notable in the light of Lashley's rhetoric of neutrality.

The committee's emphasis on pure science detached from the eugenists' social applications was a subtle turning point in the history of genetics. It marks a change from the eugenist concern with social amelioration and from the traditional progressive eugenic ideology, toward a conservative belief in the immobility of social classes. Traditional eugenists believed that society could be altered for the better, that genius needed to be cultivated and the feebleminded discouraged from breeding. But the Committee on Human Heredity, with its twin emphases on the hereditary determination of mental traits and on pure science divorced from social implications, affirmed the organization and stratification of society. This subtle shift from progressive to conservative science also took place when the Yerkes Laboratories came under Lashley's direction, and it is a theme familiar from his controversy with Herrick.

The work of the geneticist John Paul Scott also parallels Lashley's interests in important ways. There were both ideological and personal connections between Lashley and Scott. Both held Ph.D.s in genetics and were interested in the behavior of mammals. Lashley centered his energies on basic research with an ultimately conservative purpose in mind; Scott was directed to do the same by his major supporter, Alan Gregg of the Rockefeller Foundation.[31] The two also had colleagues in common.

Scott's program of research at the Roscoe B. Jackson Memorial Laboratory in Bar Harbor, Maine, was called "behavior genetics," and it involved the breeding of an intelligent and good-tempered dog. It was a direct descendant of Stockard's

[29] Wright to L. H. Snyder, Nov. 11, 1940, and Lashley to Snyder, Nov. 18, 1940, Lashley Papers, Folder "Committee on Human Heredity," Yerkes.

[30] Neel to Snyder, undated letter and proposal, Lashley Papers, Folder "Committee on Human Heredity," Yerkes.

[31] Diane Paul, "The Rockefeller Foundation and the Origin of Behavior Genetics," in *The Expansion of American Biology,* 270.

dog-breeding experiments; though of course in his attempts at purity, Scott avoided Stockard's explicit racism. Funded by Gregg, director of the Rockefeller's Medical Sciences Division, Scott's dog-breeding project was part of the foundation's broader initiative to support "basic research." During the 1940s, the Rockefeller moved to provide funding for what its officers considered pure science: the goal was knowledge, without concern for its immediate practical applications.[32] As Raymond Fosdick wrote in 1952, "[a]ll this activity, whether in classical biology or the new fields on the borderline, has been in pure research – the clean, clear urge to gain new knowledge, the unimpeded reach of imaginative scholarship even when its results seem remote and unrelated."[33] In the case of behavior genetics, the research was supposed to be distinct from eugenics, with its goal of utilitarianism and emphasis on social amelioration.

Along with their rejection of the eugenists' practical goals, the behavior geneticists and their supporters also jettisoned the eugenist progressive agenda. Alan Gregg, for example, believed that if Scott's breeding experiments accomplished their goal, they would show that mental traits such as intelligence and temperament were indeed carried on the genes. For Gregg, the purpose of breeding a smart dog was not to further the eugenic ideal of breeding a better human race. Rather, the experiments would provide proof of the innate basis of intelligence, and thereby show the ineffectiveness of mass education. Instead of demonstrating how radically society could change, as the eugenists hoped to do, Gregg wanted his Rockefeller-sponsored researchers to show how it must remain entirely the same.[34]

Yet even as the Rockefeller officers jettisoned eugenics, as Diane Paul has noted, the science of behavior genetics bore distinctly eugenic markings.[35] First, like the eugenists, behavior geneticists were interested in mentality and mental inheritance, considering physical characteristics less important, and the conse-

[32] On the Rockefeller Foundation and its support of molecular biology in particular, see Robert E. Kohler, "The Management of Science: The Experience of Warren Weaver and the Rockefeller Foundation Program in Molecular Biology," *Minerva* 14 (1976): 279–306; and Pnina Abir-Am, "The Discourse of Physical Power and Biological Knowledge in the 1930s: A Re-appraisal of the Rockefeller Foundation's 'Policy' in Molecular Biology," *Social Studies of Science* 12 (1982): 341–382. In *The Molecular Vision of Life: Caltech, the Rockefeller Foundation and the Rise of the New Biology* (New York: Oxford, 1993), Lily Kay argues that the emphasis on "purity" by Rockefeller-supported molecular biologists actually belied their interest in social control. See also Ellen Condliffe Lagemann, *The Politics of Knowledge: The Carnegie Corporation, Philanthropy and Public Policy* (Middletown: Wesleyan University Press, 1989).

[33] Raymond B. Fosdick, *The Story of the Rockefeller Foundation* (New York: Harper and Brothers, 1952), 164.

[34] On Alan Gregg and his agenda for behavior genetics, see Diane Paul, "The Rockefeller Foundation and the Origin of Behavior Genetics" in *The Expansion of American Biology*. On Scott's research at the Jackson Lab, see John Paul Scott and John L. Fuller, "Research on Genetics and Social Behavior at the Roscoe B. Jackson Memorial Laboratory, 1946–1951: A Progress Report," *Journal of Heredity* 42 (1951): 191–6.

[35] Paul, 263–4.

quences of mental development anyway. Lashley's concern was, of course, also with the hereditary determination of intelligence – an interest that represented a direct link to the eugenics movement.

Second, the dog-breeding project was supposed to be an example of Rockefeller-funded "pure science," but the results of the experiment were going to be generalized to human beings, just as eugenist conclusions were. The ease of extrapolation from other animals to humans provides another link between Lashley and the behavior genetics program. Finally, Scott emphasized the importance of temperament in intelligence, and as a result sought to breed an intelligent dog that was also good-tempered. The interest in temperament and its relationship to intelligence also appeared in Lashley's work, especially in his studies of delinquency at the Behavior Research Fund.

Scott's behavior genetics, then, shows strong parallels to Lashley's work: their interests converged on the hereditary basis of mentality and behavior, despite the fact that Scott's disciplinary affiliation was to genetics, and Lashley's (formally, at least) to psychology. Their example suggests that common research agendas united biology and psychology.

But Lashley and Scott not only shared a research agenda; their spheres also intersected on a personal level. On more than one occasion, associates of Lashley came together with behavior geneticists to share the results of their work in heredity. For example, in the summer of 1946, Scott held a conference at the Jackson Lab on Genetics and Social Behavior; present were Lashley's colleague Robert Yerkes and Lashley's former student Frank Beach. By exploring these ideological and personal connections between biology and psychology, we can see that programs like behavior genetics were actually much broader in scope than they may initially appear. Diane Paul, for instance, notes that the Jackson Lab project was a failure because Scott did not prove the limits of mass education, as Gregg had hoped, and afterwards "the study was effectively buried."[36] Nonetheless, although Scott's particular project may not have come to fruition, the broader Rockefeller initiative to demonstrate the hereditary nature of intelligence was certainly supported elsewhere, largely through Lashley's influence: through his directorship of the Yerkes Laboratories, and through his membership on the Committee for Human Heredity.

[36] Ibid., 279.

Epilogue:
Lashley and American Neuropsychology

... [B]less you for saying to me what you do about his "accomplishing nothing," and putting it in the right perspective. It did bother him that, after exploding his own latest theory, expressed in the Princeton lectures, he had no alternative to offer. But I think he still felt he had pointed a way, and it was up to the next generation now.

Claire Schiller Lashley[1]

In my interpretation, Karl Lashley's life and work illustrate three basic themes that I want now to review and to tie together. The first is the idea that during the period from 1910 to 1955, the mind-body problem – the puzzle of the relationship between consciousness and matter – became inseparably involved with the nature-nurture dispute: which was the greater determinant of intelligence and ability, heredity or environment? It is difficult to find a life scientist from the first half of this century who did not express an opinion on the nature-nurture question, especially as genetics developed in scientific and social significance, and the question took on increasing urgency. As a result, understanding psychologists' and biologists' answers to the problem of mind, brain, and behavior is impossible unless those answers are seen in the light of the nature-nurture debate.

Examining the history of the mind-body problem in terms of the nature-nurture problem, as I have done here, reveals patterns in the relationships among major life scientists, patterns that had heretofore remained obscure. We can begin to understand more clearly the differences between Lashley and Watson, Herrick, and Hull, and the similarities between Lashley, Jennings, and Köhler. We can begin to see what motivated and what was at stake in their controversies about the structure and function of the brain, and the brain's relationship to mind and behavior.

But in order to see such patterns emerge, to reveal these new stories about the mind-body problem in the twentieth century, the histories of biology and psychology must be seen as interconnected. As historians, we must broaden our focus from individual disciplinary histories to the history of life science, including biology, psychology, and the social sciences. Only then will we see the full

[1] Claire Schiller Lashley to Heinrich Klüver, November 30, 1958, "Klüver" folder, Claire Schiller Lashley Papers UFG. (The original was wrongly dated.)

187

diversity of players in these debates, and perceive their relationships to one another. Questions as crucial and as fundamental as the mind-body problem cannot be contained by disciplinary borderlines.

The second theme that this study illustrates, related to the first, is the relationship between Lashley's neuropsychology and genetics. For Lashley, genetics was an ideal and an ideology, not a scientific research program. Beyond his early dissertation work with Jennings, the experiments he supervised at the Yerkes Labs, the grants he supported on the Heredity Committee, and his informal ties to behavior genetics, Lashley himself did not do sustained research in genetics. Nonetheless, genetics represented a model science for him in two respects. In one sense, genetics was a thoroughly reductive and materialistic science; in the biochemistry of chromosomes, geneticists like T. H. Morgan had found the material, physical basis for phenotypic traits, thereby exemplifying the kind of reduction to physical reality that Lashley hoped for in psychology.

In another sense as well, genetics represented an ideology for Lashley: as genetics developed during the first half of the twentieth century, it helped to reconfigure the nature-nurture issue, giving greater plausibility to hereditarian conceptions of intelligence and ability. For Lashley, who always had hereditarian leanings, genetics gave belief in the power of heredity a "tie to physiological reality."[2] Thus genetics served to strengthen Lashley's hereditarianism, and also to bolster his hope that psychology would find a material basis not only in the brain, but even ultimately in the genes. Lashley's interest in heredity, encouraged by his graduate work with Jennings, helps to explain his break from Watson and behaviorism in the 1920s. His hereditarianism, biological determinism, and genetic ideology were also the principal factors that drove him apart from Herrick and Hull. As his hereditarianism became interconnected with a desire to maintain the status quo, it drove him apart from Herrick; and as it made psychology indistinguishable from biology, it established an unbridgeable gap between him and Hull.

Finally, the third theme I have illuminated here is the inextricable interweaving of science and society. In my interpretation of Lashley, social and political commitments were inseparably bound up with his theorizing and experimentation. He was not "neutral" or "pure" in the sense that he himself maintained, nor in the way that psychologists and historians have claimed for him since his death. Rather, I have argued, his very purity, or attempts at purity, constituted a politics. I have tried to look beyond his self-proclaimed purity, and to understand the politics that it concealed.

What has Lashley's impact or influence been? To what extent have his ideas succeeded? From a perspective limited to the history of psychology, Lashley has

---

[2] Lashley, "Experimental Analysis of Instinctive Behavior," *Neuropsychology of Lashley,* 373.

left a mixed legacy. On the one hand, his experimental method – brain ablation and the training and testing of animals – has certainly been superseded by methods of single-cell recording and more recently by techniques of brain imaging. As a result of these new methods, Lashley's theories of equipotentiality and mass action, of whole-brain functioning, were questioned toward the end of his life, and have been thrown into increasing doubt since his death. A recent *New York Times* article, for example, reported that Lashley's ablation experiments were "too imprecise" to find the localization of memory and of other functions which, it is now supposed, does exist in the brain.[3]

On the other hand, Lashley's long-standing opposition to behaviorism, to its stimulus-response and connectionist conceptions of learning, has fared somewhat better. His emphasis on attention, on inner mental process, on memory and language, has been borne out by the cognitive sciences, whose founders, as I noted in the opening of this book, revered Lashley as a hero. In many ways, of course, his approach deviated from that of the cognitive sciences. In particular, his rejection of the machine analogy for the mind and his interest in the biological correlates of intelligence, were entirely foreign to the cognitivist point of view. For cognitivists like George Miller, Jerry Fodor, and Herbert Simon, the concept of artificial intelligence was a central tenet of their new science of mind; as was the idea of a level of "mental function," subject to its own laws and independent of both biology and culture.[4] But cognitive scientists found a predecessor in Lashley because of his relentless attacks on behaviorism, because of his turn away from environmental influence in shaping behavior, and because of his exploration of the animal's inherent mental power. The behaviorism/cognitivism issue is still alive today in psychology, but cognitivism arguably has the upper hand.[5]

For a point of view limited to psychology, then, Lashley's impact has been of mixed significance. In a broader sense, however, from a perspective that considers the sciences of life, mind, and society more generally, Lashley's legacy of "anti-Progressivism" persists to the present day, especially in contemporary American popular culture. Almost daily we are confronted with genetic explanations for all manner of traits and propensities. Is homosexuality inherited? Is there a gene for sociability? For aggression? Are some people just congenitally unteachable? Given the increasing popularity of affirmative answers to these questions, Lashley's vision of the unity of science, of the reduction of psychology to physiology and ultimately to genetics, has a strikingly modern ring.

---

[3] Philip J. Hilts, "A Brain Unit Seen as Index for Recalling Memories," *The New York Times*, September 24, 1991, C8.

[4] Howard Gardner, *The Mind's New Science: A History of the Cognitive Revolution* (New York: Basic Books, 1985).

[5] For a contemporary discussion of these issues, see Abram Amsel, *Behaviorism, Neobehaviorism and Cognitivism in Learning Theory: Historical and Contemporary Perspectives* (Hillsdale: Erlbaum, 1989).

The story of genetic determinism, however, has been told almost entirely from the geneticists' point of view: our current picture of our genetic selves has been seen, rightly, as an extension of eugenics, now simply with the support of molecular genetics behind it. While there certainly are eugenic overtones to many of these current hereditarian arguments, they also contain a certain element that is decidedly noneugenist in flavor: an acceptance of the status quo as the natural order of things that cannot and should not be changed. If one is born with a "gay gene," there is nothing to be done about it. This kind of hereditarianism, implying a preservation of the current order, comes not out of eugenics but out of neuropsychology. The eugenic belief in breeding a better race has not simply dropped away from current hereditarian arguments; it has been actively excised by the anti-Progressive tradition in neuropsychology.

Lashley's beliefs that psychology should ultimately be subsumed by biology, that genetic factors determined both physical and mental traits, and that animal models were valid for humans, have been repeatedly reaffirmed in two areas of research: sex and intelligence. As the advocates of a "gay gene" accept a genetic model of sexuality and oppose a psychoanalytic argument about homosexuality, so Lashley spawned an entire line of scientific studies of sexual physiology and sexuality that rejected the psychoanalytic tradition. As we have seen, Lashley himself served for over thirty years on the National Research Council Committee for Research in Problems of Sex, and advised Alfred Kinsey on Kinsey's study of female sexuality. Among Lashley's students, Frank Beach worked on sexual function in rats; Calvin Stone continued this work and was a contributor to the volume *Sex and Internal Secretions;* Carney Landis wrote *Personality and Sexuality of the Handicapped Woman.* Landis in turn was the dissertation adviser of Albert Ellis, founder of the Society for the Scientific Study of Sex.[6]

The recent resurgence of popular interest in the biological basis of intelligence also represents a continuation of the Lashleyan tradition. The anti-Progressive marginalization of environmental factors, in fact, forms the structuring assumption of the most notorious of these recent discussions, *The Bell Curve,* by Richard Herrnstein and Charles Murray.[7] The authors of *The Bell Curve* rely completely on a central assumption of Lashley's school: that it is possible to achieve a uniform environment, and that as environmental factors become increasingly uniform, the genetic basis of behavior will loom ever larger. Lashley achieved such a standardized environment by raising the laboratory to a preeminent posi-

[6] Beach, "Sex Reversals in the Mating Pattern of the Rat," *Journal of Genetic Psychology* 53 (1938): 329–334; Edgar Allen, ed., *Sex and Internal Secretions: A Survey of Recent Research* (Baltimore: Williams and Wilkins, 1932); Landis and M. Marjorie Bolles, *Personality and Sexuality of the Physically Handicapped Woman* (New York: P. B. Hoeber, 1942). Ellis earned a Ph.D. in psychology at Columbia University under Carney Landis.
[7] Richard J. Herrnstein and Charles Murray, *The Bell Curve: Intelligence and Class Structure in American Life* (New York: Free Press, 1994).

tion in his science. Herrnstein and Murray believe that such a standardization has taken place in society itself; in other words, they have bypassed the process by which Lashley reconstructed society inside the laboratory and have instead begun to treat society itself as a laboratory. They then move directly to Lashley's anti-Progressive conclusion. Arguing that heritability of IQ has steadily risen during the last century, Herrnstein and Murray write:

> This last point is especially important in the modern societies, with their intense efforts to equalize opportunity. *As a general rule, as environments become more uniform, heritability rises.* When heritability rises, children resemble their parents more, and siblings increasingly resemble each other; in general family members become more similar to each other and more different from people in other families. It is the central irony of egalitarianism: uniformity in society makes the members of families more similar to each other and members of different families more different. [emphasis in original][8]

This is the basic assumption on which the entire argument rests, yet the book's critics have hardly given it a second thought. Their criticisms have depended instead on unearthing the "tainted sources" of *The Bell Curve,* on connecting it with eugenics and Nazism – on showing, in other words, that it is bad science. Yet its major assumption comes out of a tradition that is most certainly good science: a tradition that without doubt helped to elevate psychology to scientific status, one distinguished for its seeming neutrality. *The Bell Curve* does not come from the fringe. The assumptions it embodies about heredity and environment and their roles in determining behavior – assumptions that undergird everything its authors say later about race and intelligence – lie deep in the most mainstream elements of psychology and biology; they share in a tradition that also fed Lashley's neuropsychology. This relationship suggests a closer connection between *The Bell Curve* and these more mainstream components than has hitherto been acknowledged.

This does not mean that therefore *The Bell Curve* is immune from criticism. To the contrary: the uncovering of this assumption makes its argument subject to criticism at an even more fundamental level. Society is not a laboratory. Herrnstein and Murray cannot possibly prove that our society presents a "uniform" environment to everyone. They have no way of demonstrating that the so-called level playing field actually exists. Their argument rests on an unquestioned belief in the myth of equal opportunity.

These examples show how the assumptions of Lashley's neuropsychology have become absorbed into popular current biological determinism. His approach ultimately has triumphed, and to such an extent that it is difficult for us today to take seriously the possibility that psychology is not reducible to biology, that animal

---

[8] Ibid., 106.

analogies provide little information when it comes to understanding how the human mind works, or that the laboratory is an inappropriate place for investigating the human psyche. Lashley's redrawing of the boundaries between biology and psychology is now taken so completely for granted that his methods and assumptions have become transparent to us; the neutrality that he always claimed for himself has become a self-fulfilling prophecy. But before we conclude that there are no alternatives to his vision, we must realize that biological determinism is a double-edged sword that most certainly cuts both ways: for every supposedly emancipatory project, such as the "gay gene," that it inspires, there is a retrograde study of race and IQ.

In the end we have to realize that neuropsychology cannot provide us with answers to our social problems; our biology cannot determine who we are. Rather, the role that scientific knowledge should play in shaping social policy will become clear only when the screen of "purity" that shields much contemporary work in the life sciences begins to be dismantled, and the political assumptions behind that work stand revealed. Only then will we, in this culture saturated with science and technology, cease to revere scientists as social authorities and to take their word as law, and begin to understand the kind of guidance that scientific expertise can provide.

Appendix: Archives Holding Lashley Material

Archives of the History of American Psychology, Bierce Library, University of Akron, Akron, Ohio.
This is an extremely well-organized and well-catalogued archive, and Lashley appears both as correspondent and as subject in a number of collections of papers. The richest material, however, is in the Kenneth W. Spence Papers, which contains correspondence – dozens of letters in all – between Lashley, Spence, and Clark L. Hull; and correspondence about Lashley between Spence and Hull. The archive also holds Lashley's responses to a questionnaire sent to eminent scientists by Lauren Wispe.

Archives of the Yerkes Regional Primate Research Center, Emory University, Atlanta, Georgia.
The collection of Lashley materials here includes mainly unpublished writings from the period of his directorship of the laboratory, most notably the annual reports of the director. There is also a great deal of material relating to Lashley's participation in various scientific societies, including the National Academy of Sciences, and government-sponsored scientific committees, including the Committee on Sensory Devices, the Committee on the National Science Foundation, and the Committee on Human Heredity. The archive also holds many of the books that were in Lashley's own library at his death.

The Alan Mason Chesney Medical Archives, The Johns Hopkins Medical Institutions, Baltimore, Maryland.
The voluminous, well-catalogued Adolf Meyer Papers contain correspondence between Meyer and Lashley, Robert M. Yerkes, Knight Dunlap, C. Judson Herrick, Clarence Luther Herrick, H. S. Jennings, and L. L. Thurstone.

Archives of the University of Florida at Gainesville.
This archive contains the remarkable 1950s correspondence between Lashley and John B. Watson, as well as the letters of Lashley and George H. Bishop, also from the 1950s. There are also smaller collections of letters between Lashley and members of his family, and many of Lashley's unpublished typescripts for lec-

tures. This material, all from roughly the last fifteen years of Lashley's life, was given to the university by Christina Schiller Schlusemeyer, Lashley's stepdaughter. The Paul Schiller papers, which this archive also holds, contain some material regarding Lashley.

The American Philosophical Society Library, Philadelphia, Pennsylvania.
Lashley materials are scattered throughout several of the Library's collections, including the papers of A. F. Blakeslee, Wolfgang Köhler, Franz Boas, Herbert Spencer Jennings, Warren Sturgis McCulloch, and Raymond Pearl. But the most significant items are contained in the Leonard Carmichael Papers and the Anne Roe Papers; the latter includes the lengthy transcript of an interview that Roe conducted with Lashley in 1950.

Bibliography

Published Works by Karl Spencer Lashley (1912–1957)

Thirty-one of Lashley's papers have been collected in *The Neuropsychology of Lashley: Selected Papers of K. S. Lashley.* eds. Frank A. Beach, Donald O. Hebb, Clifford T. Morgan, and Henry W. Nissen. New York: McGraw Hill, 1960.

From the Johns Hopkins University
Lashley, K. S. "Visual Discrimination of Size and Form in the Albino Rat." *Journal of Animal Behavior* 2 (1912): 310–331.
Jennings, H. S. and K. S. Lashley. "Biparental Inheritance and the Question of Sexuality in Paramecium." *Journal of Experimental Zoology* 14 (1913): 393–466.
Jennings, H. S. and K. S. Lashley. "Biparental Inheritance of Size in Paramecium." *Journal of Experimental Zoology* 15 (1913): 193–200.
Lashley, K. S. and John B. Watson. "Notes on the Development of a Young Monkey." *Journal of Animal Behavior* 3 (1913): 114–139.
Lashley, K. S. "Reproduction of Inarticulate Sounds in the Parrot." *Journal of Animal Behavior* 3 (1913): 361–366.
Watson, John B. and K. S. Lashley. "Literature for 1912 on the Behavior of Vertebrates." *Journal of Animal Behavior* 3 (1913): 446–463.
Lashley, K. S. "A Note on the Persistence of an Instinct." *Journal of Animal Behavior* 4 (1914): 293–4.
Lashley, K. S. "Recent Literature of a General Nature on Animal Behavior." *Psychological Bulletin* 11 (1914): 269–277.
Watson, John B. and K. S. Lashley. "An Historical and Experimental Study of Homing." *Carnegie Institution Publications* 7.211 (1915): 9–60.
Lashley, K. S. "Notes on the Nesting Activities of the Noddy and Sooty Terns." *Carnegie Institution Publications* 7.211 (1915): 61–83.
Lashley, K. S. "The Acquisition of Skill in Archery." *Carnegie Institution Publications* 7.211 (1915): 107–128.
Lashley, K. S. "Inheritance in the Asexual Reproduction of Hydra viridis." *Proceedings of the National Academy of Sciences, Washington, D.C.* 1 (1915): 298–301.
Lashley, K. S. "Recent Literature on Sensory Discrimination in Animals." *Psychological Bulletin* 12 (1915): 219–299.
Lashley, K. S. "Inheritance in the Asexual Reproduction of Hydra." *Journal of Experimental Zoology* 19 (1915): 157–210.
Lashley, K. S. "Results of Continued Selection in Hydra." *Journal of Experimental Zoology* 20 (1916): 19–26.
Mast, Samuel O. and K. S. Lashley. "Observations on Ciliary Current in Free-Swimming Paramecia." *Journal of Experimental Zoology* 21 (1916): 281–293.

Lashley, K. S. "The Color Vision of Birds I: The Spectrum of the Domestic Fowl." *Journal of Animal Behavior* 6 (1916): 1–26.

Lashley, K. S. "Sensory Physiology of Animals." *Psychological Bulletin* 13 (1916): 309–315.

Lashley, K. S. "The Human Salivary Reflex and Its Use in Psychology." *Psychological Review* 23 (1916): 446–464.

Lashley, K. S. "Reflex Secretion of the Human Parotid Gland." *Journal of Experimental Psychology* 1 (1916): 461–493.

Lashley, K. S. "Changes in the Amount of Salivary Secretion Associated With Cerebral Lesions." *American Journal of Physiology* 43 (1917): 62–72.

Lashley, K. S. "The Accuracy of Movement in the Absence of Excitation from the Moving Organ." *American Journal of Physiology* 43 (1917): 169–194.

Franz, Shepherd Ivory and K. S. Lashley. "The Retention of Habits by the Rat after Destruction of the Frontal Portions of the Cerebrum." *Psychobiology* 1 (1917): 3–18.

Lashley, K. S. "Sensory Physiology of Animals." *Psychological Bulletin* 14 (1917): 276–283.

Lashley, K. S. and Shepherd Ivory Franz. "The Effects of Cerebral Destruction upon Habit Formation and Retention in the Albino Rat." *Psychobiology* 1 (1917): 71–140.

Lashley, K. S. "The Effects of Strychnine and Caffeine upon the Rate of Learning." *Psychobiology* 1 (1917): 141–170.

Lashley, K. S. "The Criterion of Learning in Experiments with the Maze." *Journal of Animal Behavior* 7 (1917): 66–70.

Hubbert, H. B. and K. S. Lashley. "Retroactive Association and the Elimination of Errors in the Maze." *Journal of Animal Behavior* 7 (1917): 130–138.

Lashley, K. S. "A Causal Factor in the Relation of the Distribution of Practice to the Rate of Learning." *Journal of Animal Behavior* 7 (1917): 139–142.

Lashley, K. S. "Modifiability of the Preferential Use of the Hands in the Rhesus Monkey." *Journal of Animal Behavior* 7 (1917): 178–186.

Lashley, K. S. "A Simple Maze: with Data on the Relation of the Distribution of Practice to the Rate of Learning." *Psychobiology* 1 (1918): 353–367.

Lashley, K. S. and J. D. Dodson. "Sensory Physiology of Animals." *Psychological Bulletin* 16 (1919): 159–164.

Lashley, K. S. "Studies of Cerebral Function in Learning." *Psychobiology* 2 (1920): 55–135.

Watson, John B. and K. S. Lashley. "A Consensus of Medical Opinion upon Questions Relating to Sex Education and Venereal Disease Campaigns." *Mental Hygiene* 4 (1920): 769–847.

From the University of Minnesota

Lashley, K. S. "Sensory Physiology of Animals." *Psychological Bulletin* 17 (1920): 178–187.

Lashley, K. S. and John B. Watson. "A Psychological Study of Motion Pictures in Relation to Venereal Disease Campaigns." *Publications of the United States Interdepartmental Social Hygiene Board* (1922): 1–88.

Lashley, K. S. "Studies of Cerebral Function in Learning II: The Effects of Long-Continued Practice upon Localization." *Journal of Comparative Psychology* 1 (1921): 453–468.

Lashley, K. S. "Studies of Cerebral Function in Learning III: The Motor Areas." *Brain* 44 (1921): 255–286.

Lashley, K. S. "Studies of Cerebral Function in Learning IV: Vicarious Function after Destruction of the Visual Areas." *American Journal of Physiology* 59 (1922): 44–71.

Lashley, K. S. "The Behavioristic Interpretation of Consciousness." *Psychological Review* 30 (1923): 237–272, 329–353.
Lashley, K. S. "Studies of Cerebral Function in Learning V: The Retention of Motor Habits after Destruction of the So-Called Motor Areas in Primates." *Archives of Neurology and Psychiatry, Chicago* 12 (1924): 249–276.
Lashley, K. S. "Physiological Analysis of the Libido." *Psychological Review* 31 (1924): 192–202.
Lashley, K. S. "Studies of Cerebral Function in Learning VI: The Theory that Synaptic Resistance is Reduced by the Passage of the Nerve Impulse." *Psychological Review* 31 (1924): 369–375.
Lashley, K. S. "Studies of Cerebral Function in Learning VII: The Relation Between Cerebral Mass, Learning and Retention." *Journal of Comparative Neurology* 41 (1926): 1–58.
Lashley, K. S. "Temporal Variation in the Function of the Gyrus Precentralis in Primates." *American Journal of Physiology* 65 (1923): 585–602.
Lashley, K. S. and D. A. McCarthy. "The Survival of the Maze Habit after Cerebellar Injuries." *Journal of Comparative Psychology* 6 (1926): 423–433.

From the Institute for Juvenile Research
Lashley, K. S. and Josephine Ball. "Spinal Conduction and Kinesthetic Sensitivity in the Maze Habit." *Journal of Comparative Psychology* 9 (1929): 71–105.
Lashley, K. S. *Brain Mechanisms and Intelligence: A Quantitative Study of Injuries to the Brain.* Chicago: University of Chicago Press, 1929.
Lashley, K. S. "Learning I: Nervous Mechanisms in Learning." *The Foundations of Experimental Psychology,* ed. Carl Murchison. Worcester: Clark University Press, 1929, 524–563.
Lashley, K. S. "Basic Neural Mechanisms in Behavior." *Psychological Review* 37 (1930): 1–24.
Lashley, K. S. "The Mechanism of Vision I: A Method for Rapid Analysis of Pattern-Vision in the Rat." *Journal of Genetic Psychology* 37 (1930): 453–460.
Lashley, K. S. "The Mechanism of Vision II: The Influence of Cerebral Lesions upon the Threshold of Discrimination for Brightness in the Rat." *Journal of Genetic Psychology* 37 (1930): 461–480.

From the University of Chicago
Lashley, K. S. "The Mechanism of Vision III: The Comparative Visual Acuity of Pigmented and Albino Rats." *Journal of Genetic Psychology* 37 (1930): 481–484.
Lashley, K. S. "The Mechanism of Vision IV: The Cerebral Areas Necessary for Pattern Vision in the Rat." *Journal of Comparative Neurology* 53 (1931): 419–478.
Lashley, K. S. "Cerebral Control Versus Reflexology: A Reply to Professor Hunter." *Journal of General Psychology* 5 (1931): 3–20.
Lashley, K. S. "Mass Action in Cerebral Function." *Science* 73 (1931): 245–254.
Lashley, K. S. "The Mechanism of Vision V: The Structure and Image-Forming Power of the Rat's Eye." *Journal of Comparative Psychology* 13 (1932) 173–200.
Lashley, K. S and M. Frank. "The Mechanism of Vision VI: The Lateral Portion of the Area Striata in the Rat: a Correction." *Journal of Comparative Neurology* 55 (1932): 525–529.
Lashley, K. S. "Studies of Cerebral Function in Learning VIII: A Reanalysis of Data on Mass Action in the Visual Cortex." *Journal of Comparative Neurology* 54 (1932): 77–84.

Lashley, K. S. "Massenleistung und Gehirnfunktionen." *Nervenarzt* 5.3 (1932): 113–120, 180–184.
Lashley, K. S. Introduction. *Studies in the Dynamics of Behavior,* ed. Lashley. Chicago: University of Chicago Press, 1932, vii–x.
Lashley, K. S. and L. E. Wiley. "Studies of Cerebral Function in Learning IX: Mass Action in Relation to the Number of Elements in the Problem to be Learned." *Journal of Comparative Neurology* 57 (1933): 3–56.
Lashley, K. S., W. T. McDonald and H. N. Peters. "Studies of Cerebral Function in Learning X: The Effects of Dilatation of the Ventricles upon Maze-Learning." *American Journal of Physiology* 104 (1933): 51–61.
Lashley, K. S. "Integrative Functions of the Cerebral Cortex." *Physiological Review* 13 (1933): 1–42.
Lashley, K. S. Introduction. *Behavior Mechanisms in Monkeys,* by Heinrich Klüver. Chicago: University of Chicago Press, 1933, ix–x.
Lashley, K. S. "Learning III: Nervous Mechanisms in Learning." *A Handbook of General Experimental Psychology,* ed. Carl Murchison. Worcester: Clark University Press, 1934, 456–496.
Lashley, K. S. "The Mechanism of Vision VII: The Projection of the Retina upon the Primary Optic Centers in the Rat." *Journal of Comparative Neurology* 59 (1934): 341–373.
Lashley, K. S. "The Mechanism of Vision VIII: The Projection of the Retina upon the Cerebral Cortex of the Rat." *Journal of Comparative Neurology* 60 (1934): 57–79.
Lashley, K. S. and M. Frank. "The Mechanism of Vision X: Postoperative Disturbances of Habits Based on Detail Vision in the Rat after Lesions in the Cerebral Visual Areas." *Journal of Comparative Psychology* 17 (1934): 355–391.
Lashley, K. S. and J. T. Russell. "The Mechanism of Vision XI: A Preliminary Test of Innate Organization." *Journal of Genetic Psychology* 45 (1934): 136–144.
Lashley, K. S. "Studies of Cerebral Function in Learning XI: The Behavior of the Rat in Latch-Box Situations." *Comparative Psychology Monographs* 11 (1935): 1–42.
Lashley, K. S. "The Mechanism of Vision XII: Nervous Structures Concerned in the Acquisition and Retention of Habits Based on Reactions to Light." *Comparative Psychology Monographs* 11.52 (1935): 43–79.

From Harvard University
Lashley, K. S. "The Mechanism of Vision XIII: Cerebral Function in Discrimination of Brightness when Detail Vision is Controlled." *Journal of Comparative Neurology* 66 (1937): 471–480.
Lashley, K. S. "The Mechanism of Vision XIV: Visual Perception of Distance after Injuries to the Cerebral Cortex, Colliculi or Optic Thalamus." *Journal of Genetic Psychology* 51 (1937): 187–207.
Lashley, K. S. "Functional Determinants of Cerebral Localization." *Archives of Neurology and Psychiatry, Chicago* 38 (1937): 371–387.
Lashley, K. S. "The Thalamus and Emotion." *Psychological Review* 45 (1938): 42–61.
Lashley, K. S. "The Mechanism of Vision XV: Preliminary Studies of the Rat's Capacity for Detail Vision." *Journal of General Psychology* 18 (1938): 123–193.
Lashley, K. S. "Conditional Reactions in the Rat." *Journal of Psychology* 6 (1938): 311–324.
Lashley, K. S. "Experimental Analysis of Instinctive Behavior." *Psychological Review* 45 (1938): 445–471.

Lashley, K. S. "Factors Limiting Recovery after Central Nervous Lesions." *Journal of Nervous and Mental Diseases* 88 (1938): 733–755.

Lashley, K. S. "The Mechanism of Vision XVI: The Functioning of Small Remnants of the Visual Cortex." *Journal of Comparative Neurology* 70 (1939): 45–67.

Lashley, K. S. Foreword. *The Organism: A Holistic Approach to Biology Based on Pathological Data in Man,* by Kurt Goldstein. Boston: Beacon Press, 1963, (reprint of 1939 edition), xii–xiii.

Lashley, K. S. "Studies of Simian Intelligence from the University of Liege." *Psychological Bulletin* 37 (1940): 237–248.

Lashley, K. S. "Coalescence of Neurology and Psychology." *Proceedings of the American Philosophical Society* 84 (1941): 461–470.

Lashley, K. S. "Patterns of Cerebral Integration Indicated by the Scotomas of Migraine." *Archives of Neurology and Psychiatry, Chicago* 46 (1941): 331–339.

Lashley, K. S. "Thalamo-Cortical Connections of the Rat's Brain." *Journal of Comparative Neurology* 75 (1941): 67–121.

Lashley, K. S. "An Examination of the 'Continuity Theory' as Applied to Discriminative Learning." *Journal of General Psychology* 26 (1942): 241–265.

Lashley, K. S. "The Mechanism of Vision XVII: Autonomy of the Visual Cortex." *Journal of Genetic Psychology* 60 (1942): 197–221.

Lashley, K. S. "The Problem of Cerebral Organization in Vision." *Biological Symposia* 7 (1942): 301–322.

From the Yerkes Laboratories of Primate Biology

Lashley, K. S and Roger W. Sperry. "Olfactory Discrimination after Destruction of the Anterior Thalamic Nuclei." *American Journal of Physiology* 139 (1943): 446–450.

Lashley, K. S. "Studies of Cerebral Function in Learning XII: Loss of the Maze Habit after Occipital Lesions in Blind Rats." *Journal of Comparative Neurology* 79 (1943): 431–462.

Lashley, K. S. "Studies of Cerebral Function in Learning XIII: Apparent Absence of Transcortical Association in Maze Learning." *Journal of Comparative Neurology* 80 (1944): 257–281.

Lashley, K. S. "Sensory Control and Rate of Learning in the Maze." *Journal of Genetic Psychology* 66 (1945): 143–145.

Lashley, K. S. and M. Wade. "The Pavlovian Theory of Generalization." *Psychological Review* 53 (1946): 72–87.

Lashley, K. S. and G. Clark. "The Cytoarchitecture of the Cerebral Cortex of Ateles: A Critical Examination of Architectonic Studies." *Journal of Comparative Neurology* 85 (1946): 223–306.

Lashley, K. S. "Structural Variation in the Nervous System in Relation to Behavior." *Psychological Review* 54 (1947): 325–334.

Lashley, K. S. "The Mechanism of Vision XVIII: Effects of Destroying the Visual 'Associative Areas' of the Monkey." *Genetic Psychology Monographs* 37.2 (1948): 107–166.

Lashley, K. S. "Persistent Problems in the Evolution of Mind." *Quarterly Review of Biology* 24 (1949): 28–42.

Lashley, K. S. "The Problem of Interaction of Cerebral Areas." *Transactions of the American Neurological Association* (1949): 187–194.

Lashley, K. S. "Psychological Problems in the Development of Instrumental Aids for the Blind." *Blindness: Modern Approaches to the Unseen Environment,* ed. Paul Zahl. Princeton: Princeton University Press, 1950, 495–511.

Lashley, K. S. "In Search of the Engram." *Symposia of the Society for Experimental Biology* 4 (1950): 454–482.

Lashley, K. S. "Physiological Psychology." *Encyclopedia Britannica,* 1950 ed.

Lashley, K. S., K. L. Chow and J. Semmes. "An Examination of the Electrical Field Theory of Cerebral Integration." *Psychological Review* 58 (1951): 123–136.

Lashley, K. S. "The Problem of Serial Order in Behavior." *Cerebral Mechanisms in Behavior,* ed. L. A. Jeffress. New York: Wiley, 1951, 112–136.

Lashley, K. S. "Neuropsychology." *Survey of Neurobiology.* Washington, D. C.: National Research Council, 1952, 18–23.

Lashley, K. S. "Dynamic Processes in Perception." *Brain Mechanisms and Consciousness,* eds. E. D. Adrian, F. Bremer, and H. H. Jasper. Springfield, IL: Charles C. Thomas, 1954, 422–443.

Lashley, K. S. "Instinct." *Encyclopedia Britannica,* 1956 ed.

Lashley, K. S. and Kenneth M. Colby. "An Exchange of Views on Psychic Energy and Psychoanalysis." *Behavioral Science* 2 (1957): 231–240.

Lashley, K. S. Introduction. *Instinctive Behavior: The Development of a Modern Concept,* trans. and ed. Claire H. Schiller. (Contributors D. J. Kuenen, Konrad Lorenz, Nicholas Tinbergen, Paul H. Schiller, Jacob von Uexküll.) New York: International Universities Press, 1957, ix–xii.

Lashley, K. S. "Cerebral Organization and Behavior." *The Brain and Human Behavior* 36 (1958): 1–18.

Other Published Sources

Aberle, Sophie D. and George W. Corner. *Twenty-Five Years of Sex Research: History of the National Research Council Committee for Research in Problems of Sex, 1922–1947.* Philadelphia: W. B. Saunders, 1953.

Abir-Am, Pnina. "The Discourse of Physical Power and Biological Knowledge in the 1930s: A Reappraisal of the Rockefeller Foundation's 'Policy' in Molecular Biology." *Social Studies of Science* 12 (1982): 341–382.

Addams, Jane. *The Second Twenty Years at Hull House: Sept. 1909–Sept. 1929.* New York: MacMillan, 1930.

Addams, Jane et al. *The Child, the Clinic and the Court.* New York: New Republic, 1927.

Adrian, E. D. "Afferent Discharges to the Cerebral Cortex from Peripheral Sense Organs." *Journal of Physiology* 100 (1941): 159–91.

Allen, Edgar., ed. *Sex and Internal Secretions: A Survey of Recent Research.* Baltimore: Williams and Wilkins, 1932.

Allen, Garland. *Life Science in the Twentieth Century.* Cambridge: Cambridge University Press, 1978.

Allport, Floyd. *Social Psychology.* Boston: Houghton Mifflin, 1924.

Amsel, Abram. *Behaviorism, Neobehaviorism and Cognitivism in Learning Theory: Historical and Contemporary Perspectives.* Hillsdale: Erlbaum, 1989.

Baernstein, H. D. and Clark Hull. "A Mechanical Parallel to the Conditioned Reflex." *Science* 70 (1929): 14–15.

Baernstein, H. D. and Clark L. Hull. "A Mechanical Model of the Conditioned Reflex." *Journal of General Psychology* 5 (1931): 99–106.

Bakan, David. "Behaviorism and American Urbanization." *Journal of the History of the Behavioral Sciences* 2 (1966): 5–28.

Bannister, Robert C. *Sociology and Scientism: The American Quest for Objectivity, 1880–1940.* Chapel Hill: University of North Carolina Press, 1987.

Barkan, Elazar. *The Retreat of Scientific Racism: Changing Concepts of Race in Britain and the United States Between the World Wars.* Cambridge: Cambridge University Press, 1992.

Bartelmez, George W. "Charles Judson Herrick, 1868–1960." *Biographical Memoirs of the National Academy of Sciences* 43 (1973): 77–108.

Beach, Frank A. "Clark Leonard Hull." *Biographical Memoirs of the National Academy of Sciences* 33 (1959): 125–141.

Beach, Frank A. "Evolutionary Changes in the Physiological Control of Mating Behavior in Mammals." *Psychological Review* 54 (1947): 297–315.

Beach, Frank A. "Karl Spencer Lashley, June 7, 1890–August 7, 1958." *Biographical Memoirs of the National Academy of Sciences* 35 (1961): 162–204.

Beach, Frank A. "Neural Basis of Innate Behavior I: Effects of Cortical Lesions Upon Maternal Behavior in the Rat." *Journal of Comparative Psychology* 24 (1937): 393–439.

Beach, Frank A. "Neural Basis of Innate Behavior II: Relative Effects of Partial Decortication in Adulthood and Infancy Upon the Maternal Behavior of the Primiparous Rat." *Journal of Genetic Psychology* 53 (1938): 109–148.

Beach, Frank A. "Sex Reversals in the Mating Pattern of the Rat." *Journal of Genetic Psychology* 53 (1938): 329–334.

Beniger, James R. *The Control Revolution: Technological and Economic Origins of the Information Society.* Cambridge: Harvard University Press, 1986.

Bennett, George K. and Lewis B. Ward. "A Model of the Synthesis of Conditioned Reflexes." *American Journal of Psychology* 45 (1933): 339–342.

Benson, Keith, Jane Maienschein and Ronald Rainger. *The Expansion of American Biology.* New Brunswick: Rutgers University Press, 1991.

Blustein, Bonnie Ellen. "Medicine as Biology: Neuropsychiatry at the University of Chicago, 1928–1939." *Perspectives on Science* 1 (1993): 416–444.

Blustein, Bonnie Ellen. "Percival Bailey and Neurology at the University of Chicago, 1928–1939." *Bulletin of the History of Medicine* 66 (1984): 90–113.

Boakes, Robert. *From Darwin to Behaviorism: Psychology and the Minds of Animals.* Cambridge: Cambridge University Press, 1984.

Boring, Edwin G. "Lashley and Cortical Integration." *The Neuropsychology of Lashley,* eds. Beach, Hebb, Morgan, and Nissen. New York: McGraw Hill, 1960, xi–xvii.

Boring, Edwin Garrigues. "Edwin Garrigues Boring." *A History of Psychology in Autobiography,* Vol. 4, eds. E. G. Boring, Herbert Langfeld, Heinz Werner, and Robert M. Yerkes. Worcester: Clark University Press, 1952.

Boring, Edwin Garrigues. *A History of Experimental Psychology.* New York: Appleton, 1929.

Brown, JoAnne. *The Definition of a Profession: The Authority of Metaphor in the History of Intelligence Testing, 1890–1930.* Princeton: Princeton University Press, 1992.

Brozek, Josef., ed. *Explorations in the History of Psychology in the United States.* Lewisburg: Bucknell University Press, 1983.

Bruce, Darryl. "Integrations of Lashley." *Portraits of Pioneers in Psychology,* eds. Gregory A. Kimble, Michael Wertheimer, and Charlotte White. Hillsdale: Erlbaum, 1991, 306–323.

Bruce, Darryl. "Lashley's Shift from Bacteriology to Neuropsychology, 1910–1917, and the Influence of Jennings, Watson and Franz." *Journal of the History of the Behavioral Sciences* 22 (1986): 27–44.

Bruce, Darryl. "On the Origin of the Term 'Neuropsychology.'" *Neuropsychologia* 23.6 (1985): 813–814.

Buckley, Kerry. *Mechanical Man: John Broadus Watson and the Beginnings of Behaviorism.* New York: Guilford Press, 1989.

Bugos, Glenn E. "Managing Cooperative Research and Borderland Science in the National Research Council, 1922–1942." *Historical Studies in the Physical and Biological Sciences* 20 (1989): 1–32.

Bullough, Vern L. "Katharine Bement Davis, Sex Research and the Rockefeller Foundation." *Bulletin of the History of Medicine* 62 (1988): 74–89.

Bullough, Vern L. "The Rockefellers and Sex Research." *Journal of Sex Research* 21 (1985): 113–125.

Burnham, John Chynoweth. "On the Origins of Behaviorism." *Journal of the History of the Behavioral Sciences* 4 (1968): 143–151.

Burnham, John Chynoweth. "Psychiatry, Psychology and the Progressive Movement." *American Quarterly* 12 (1960): 457–65.

Burnham, John Chynoweth. "The New Psychology: From Narcissism to Social Control." *Change and Continuity in Twentieth Century America: The 1920s,* eds. John Braeman, Robert H. Bremner, and David Brody. Columbus: Ohio State University Press, 1968, 351–398.

Burnham, John Chynoweth. "Thorndike's Puzzle Boxes." *Journal of the History of the Behvaioral Sciences* (April 1972): 159–167.

Burnham, John Chynoweth. *Paths into American Culture: Psychology, Medicine and Morals.* Philadephia: Temple University Press, 1988.

Camfield, Thomas. "The Professionalization of American Psychology." *Journal of the History of the Behavioral Sciences* 9 (1973): 66–75.

Carmichael, Leonard. "The Growth of the Sensory Control of Behavior Before Birth." *Psychological Review* 54 (1947): 316–324.

Carson, John. "Army Alpha, Army Brass, and the Search for Army Intelligence." *Isis* 84 (1993): 278–309.

Chatfield, Charles. *For Peace and Justice: Pacifism in America, 1910–1941.* Knoxville: University of Tennessee Press, 1971.

Child, Charles Manning. "Driesch's Harmonic Equipotential Systems in Form-Regulation." *Biologisches Centralblatt* 28 (1908): 527–623.

Child, Charles Manning. *Individuality in Organisms.* Chicago: University of Chicago Press, 1915.

Child, Charles Manning. *Origin and Development of the Nervous System from a Physiological Viewpoint.* Chicago: University of Chicago Press, 1921.

Child, Charles Manning. *Physiological Foundations of Behavior.* New York: Henry Holt, 1924.

Child, Charles Manning. *Senescence and Rejuvenescence.* Chicago: University of Chicago Press, 1915.

Clarke, Adele. "Controversy and the Development of Reproductive Sciences." *Social Problems* 37 (1990): 18–37.

Clarke, Adele. "Money, Sex and Legitimacy at Chicago, circa 1892–1940: Lillie's Center of Reproductive Biology." *Perspectives on Science* 1 (1993): 367–415.

Cobb, Stanley. "A Salute from Neurologists." *The Neuropsychology of Lashley,* eds. Beach, Hebb, Morgan, and Nissen. New York: McGraw Hill, 1960, xvii–xx.

Coghill, George Ellett. "The Structural Basis of the Integration of Behavior." *Proceedings of the National Academy of Sciences* 16 (1930): 637–43.

Coghill, George Ellett. *Anatomy and the Problem of Behavior.* New York: MacMillan, 1929.

Cohen, David. *J. B. Watson, the Founder of Behaviorism.* London: Routledge and Kegan Paul, 1979.

Cravens, Hamilton. *The Triumph of Evolution: American Scientists and the Heredity-Environment Controversy.* Philadelphia: University of Pennsylvania Press, 1978.

Cravens, Hamilton and John C. Burnham. "Psychology and Evolutionary Naturalism in American Thought." *American Quarterly* 23 (1971): 635–57.

Crosby, Elizabeth C. "Charles Judson Herrick." *Journal of Comparative Neurology* 115 (1960): 1–8.

Cross, Stephen J. and William R. Albury. "Walter B. Cannon, L. J. Henderson and the Organic Analogy." *Osiris,* second series 3 (1987): 165–192.

Danziger, Kurt. *Constructing the Subject: Historical Origins of Psychological Research.* Cambridge: Cambridge University Press, 1990.

Deegan, Mary Jo. *Jane Addams and the Men of the Chicago School.* New Brunswick: Transaction Books, 1988.

Degler, Carl N. *In Search of Human Nature: The Decline and Reversal of Darwinism in American Social Thought.* New York: Oxford University Press, 1991.

Dennett, Daniel. *Consciousness Explained.* Boston: Little, Brown and Co., 1991.

Dewsbury, Donald A. "Contributions to the History of Psychology XCIV: The Boys of Summer at the End of Summer: The Watson-Lashley Correspondence of the 1950s." *Psychological Reports* 72 (1993): 263–269.

Dewsbury, Donald A. "Psychobiology." *American Psychologist* 46.3 (1991): 198–205.

Dewsbury, Donald A. *Comparative Psychology in the Twentieth Century.* Stroudsburg: Hutchinson Ross, 1984.

Dewsbury, Donald A., ed. *Leaders in the Study of Animal Behavior.* Lewisburg: Bucknell University Press, 1985.

Dollard, John. *Personality and Psychotherapy: An Analysis in Terms of Learning, Thinking and Culture.* New York: McGraw Hill, 1950.

Donaldson, Henry Herbert. *Growth of the Brain: A Study of the Nervous System in Relation to Education.* New York: C. Scribner's, 1895.

Dummer, Ethel S. *Why I Think So: The Autobiography of an Hypothesis.* Chicago: Clarke-McElroy, 1936.

Dunlap, Knight. "Psychological Hypotheses Concerning the Functions of the Brain." *Scientific Monthly* 31 (1930): 97–112.

Dunlap, Knight. *Habits: Their Making and Unmaking.* New York: Liveright, 1932.

Dunlap, Knight. *Outline of Psychobiology.* Baltimore: The Johns Hopkins University Press, 1914.

Dunlap, Knight. *Personal Adjustment.* New York: McGraw Hill, 1946.

Dunlap, Knight. *Psychologies of 1925.* Worcester: Clark University Press, 1927.

Dunn, L. C. and Theodosius Dobzhansky. *Heredity, Race and Society.* New York: The New American Library, 1946.

Ellson, Douglas G. "A Mechanical Synthesis of Trial-and-Error Learning." *Journal of General Psychology* 13 (1935): 212–218.

Fearing, Franklin. *Reflex Action: A Study in the History of Physiological Psychology.* Baltimore: Williams and Wilkins, 1930.

Fee, Elizabeth. *Disease and Discovery. A History of the Johns Hopkins School of Hygiene and Public Health.* Baltimore: Johns Hopkins University Press, 1987.

Finger, Stanley. *Origins of Neuroscience: A History of Explorations into Brain Function.* New York: Oxford University Press, 1993.

Finison, Lorenz. "Unemployment, Politics and the History of Organized Psychology." *American Psychologist* 31 (1976): 747–755.

Fosdick, Raymond. *The Story of the Rockefeller Foundation.* New York: Harper and Brothers, 1952.

Franz, Shepherd Ivory. *Nervous and Mental Re-Education.* New York: MacMillan, 1923.

Fredrickson, George M. *The Black Image in the White Mind: The Debate on Afro-American Character and Destiny, 1817–1914.* New York: Harper and Row, 1971.

Furner, Mary O. *Advocacy and Objectivity: A Crisis in the Professionalization of American Social Science, 1865–1905.* Lexington: University Press of Kentucky, 1975.

Gardner, Howard. *The Mind's New Science: A History of the Cognitive Revolution.* New York: Basic Books, 1985.

Geiger, Roger L. *To Advance Knowledge: The Growth of American Research Universities, 1900–1940.* New York: Oxford University Press, 1986.

Geison, Gerald L., ed. *Physiology in the American Context, 1850–1940.* Bethesda: American Physiological Society, 1987.

Gesell, Arnold. "Scientific Approaches to the Study of the Human Mind." *Science* 88 (1938): 225–230.

Gesell, Arnold. "The Teacher-Pupil Relationship in a Democracy." *School and Society* 51 (1940): 193–198.

Goldstein, Kurt. *The Organism: A Holistic Approach to Biology Derived from Pathological Data in Man.* New York: American Book Company, 1939.

Gould, Stephen Jay. *The Mismeasure of Man.* New York: Norton, 1981.

Griffen, Clyde. "The Progressive Ethos." *The Development of an American Culture,* second edition, eds. Stanley Coben and Lorman Ratner. New York: St. Martin's Press, 1983.

Haber, Samuel. *Efficiency and Uplift: Scientific Management in the Progressive Era, 1890–1920.* Chicago: University of Chicago Press, 1964.

Hall, Granville Stanley. *Adolescence: Its Psychology and Its Relations to Physiology, Anthropology, Sociology, Sex Crime, Religion and Education.* New York: D. Appleton, 1904.

Haller, Mark. *Eugenics: Hereditarian Attitudes in American Thought.* New Brunswick: Rutgers University Press, 1963.

Halstead, Ward C. *Brain and Intelligence: A Quantitative Study of the Frontal Lobes.* Chicago: University of Chicago Press, 1947.

Haraway, Donna J. *Primate Visions: Gender, Race and Nature in the World of Modern Science.* New York: Routledge, 1989.

Haraway, Donna J. *Simians, Cyborgs and Women: The Reinvention of Nature.* New York: Routledge, 1991.

Harrington, Anne, "A Feeling for the 'Whole': The Holistic Reaction in Neurology from the Fin de Siècle to the Interwar Years." *Fin de Siècle and Its Legacy,* eds. Mikulas Teich and Roy Porter. Cambridge: Cambridge University Press, 1990, 254–277.

Harrington, Anne. "Other 'Ways of Knowing': The Politics of Knowledge in Interwar German Brain Science." *So Human a Brain: Knowledge and Values in the Neurosciences,* ed. Harrington. Boston: Birkhauser, 1992, 229–244.

Harrington, Anne. *Medicine, Mind and the Double Brain.* Princeton: Princeton University Press, 1987.

Haskell, Thomas L. *The Emergence of Professional Social Science: The American Social Science Association and the Nineteenth-Century Crisis of Authority.* Urbana: University of Illinois Press, 1977.

Hebb, Donald O. "The Effect of Early and Late Brain Injury Upon Test Scores, and the Nature of Normal Adult Intelligence." *Proceedings of the American Philosophical Society* 85 (1942): 275–292.

Hebb, Donald O. Introduction. *Brain Mechanisms and Intelligence,* by K. S. Lashley. New York: Dover Publications, 1963, v–xiii.

Heims, Steve Joshua. "Encounter of Behavioral Sciences with Machine-Organism Analogies in the 1940s." *Journal of the History of the Behavioral Sciences* 11 (1975): 368–373.

Heims, Steve Joshua. *Constructing a Social Science for Postwar America: The Cybernetics Group, 1946–1953.* Cambridge: MIT Press, 1991.

Herman, Ellen. *The Romance of American Psychology: Political Culture in the Age of Experts.* Berkeley: University of California Press, 1995.

Herrick, Charles Judson and Elizabeth Caroline Crosby. *A Laboratory Outline of Neurology.* Philadelphia: W. B. Saunders, 1918.

Herrick, Charles Judson. "A Biologist Looks at the Profit Motive." *Social Forces* 16 (1938): 320–327.

Herrick, Charles Judson. "A Neurologist Makes Up His Mind." *Scientific Monthly* 49 (1939): 99–110.

Herrick, Charles Judson. "Behavior and Mechanism." *Social Forces* 7 (1928): 1–11.

Herrick, Charles Judson. "Biological Determinism and Human Freedom." *International Journal of Ethics* 37 (1926): 36–52.

Herrick, Charles Judson. "Clarence Luther Herrick, Pioneer Naturalist, Teacher and Psychobiologist." *Transactions of the American Philosophical Society* 45 (1955): 1–55.

Herrick, Charles Judson. "Is the Cerebral Cortex Equipotential?" *Journal of General Psychology* 13 (1935): 398–400.

Herrick, Charles Judson. "Localization of Function in the Nervous System." *Proceedings of the National Academy of Sciences* 16 (1930): 643–650.

Herrick, Charles Judson. "Mechanism and Organism." *Journal of Philosophy* 26 (1929): 589–597.

Herrick, Charles Judson. "The Evolution of Cerebral Localization Patterns." *Science* 78 (1933): 441–444.

Herrick, Charles Judson. "The Founder and Early History of the Journal." *Journal of Comparative Neurology* 74 (1941): 25–38.

Herrick, Charles Judson. "The Spiritual Life." *Journal of Religion* 8 (1928): 505–528.

Herrick, Charles Judson. *Brains of Rats and Men: A Survey of the Origin and Biological Significance of the Cerebral Cortex.* Chicago: University of Chicago Press, 1926.

Herrick, Charles Judson. *George Ellett Coghill, Naturalist and Philosopher.* Chicago: University of Chicago Press, 1949.

Herrick, Charles Judson. *Neurological Foundations of Animal Behavior.* New York: Henry Holt, 1924.

Herrick, Charles Judson. *The Evolution of Human Nature.* Austin: University of Texas Press, 1956.

Herrick, Charles Judson. *The Thinking Machine.* Chicago: University of Chicago Press, 1929.

Herrick, Clarence Luther. "Mind and Body: The Dynamic View." *Psychological Review* 11 (1904): 395–409.

Herrnstein, Richard J. and Charles Murray. *The Bell Curve: Intelligence and Class Structure in American Life.* New York: Free Press, 1994.

Hilts, Philip J. "A Brain Unit Seen as Index for Recalling Memories." *The New York Times,* Sept. 24, 1991: C1.

Hofstadter, Richard. *The Age of Reform, from Bryan to FDR.* New York, Knopf, 1955.

Hooker, Davenport. *Evidence of Prenatal Function of the Central Nervous System in Man.* New York: American Museum of Natural History, 1958.

Hooker, Davenport. *The Prenatal Origins of Behavior.* Lawrence, Kansas: University of Kansas Press, 1952.

Hull, Clark Leonard. "Clark L. Hull." *A History of Psychology in Autobiography,* Vol. 4, eds. Edwin G. Boring, Heinz Werner, Herbert S. Langfeld, and Robert M. Yerkes. Worcester: Clark University Press, 1952, 143–162.

Hull, Clark Leonard. "Knowledge and Purpose as Habit Mechanisms." *Psychological Review* 37 (1930): 511–525.

Hull, Clark Leonard. "Mind, Mechanism and Adaptive Behavior." *Psychological Review* 44 (1937): 1–32.

Hull, Clark Leonard. "Moral Values, Behaviorism and the World Crisis." *Transactions of the New York Academy of Sciences, Series II* 7.4 (February 1945): 90–94.

Hull, Clark Leonard. "Simple Trial and Error Learning: A Study in Psychological Theory." *Psychological Review* 37 (1930): 241–256.

Hull, Clark Leonard. "The Mechanism of the Assembly of Behavior Segments in Novel Combinations Suitable for Problem Solution." *Psychological Review* 42 (1935): 219–245.

Hull, Clark Leonard. "The Place of Innate Individual and Species Differences in a Natural-Science Theory of Behavior." *Psychological Review* 52 (1945): 55–60.

Hull, Clark Leonard. *Aptitude Testing.* New York: World Book Co., 1928.

Hull, Clark Leonard. *Principles of Behavior: An Introduction to Behavior Theory.* New York: D. Appleton, 1943.

Hunter, Walter S. "A Consideration of Lashley's Theory of the Equipotentiality of Cerebral Action." *Journal of General Psychology* 3 (1930): 455–468.

Hunter, Walter S. "Summary Comments on the Heredity-Environment Symposium." *Psychological Review* 54 (1947): 348–352.

Institute for Juvenile Research. *Child Guidance Procedures: Methods and Techniques Employed at the Institute for Juvenile Research.* New York: D. Appleton, 1937.

James, William. "On Some Omissions of Introspective Psychology." *Mind* 9 (January 1884): 1–26.

Jeffress, Lloyd A., ed. *Cerebral Mechanisms in Behavior: The Hixon Symposium.* New York: John Wiley, 1951.

Jennings, Herbert Spencer, Adolf Meyer, William I. Thomas, and John B. Watson. *Suggestions of Modern Science Concerning Education.* New York: MacMillan, 1917.

Jennings, Herbert Spencer, Charles A. Berger, Dom Thomas Verner Moore, Ales Hrdlicka, Robert H. Lowie, and Otto Klineberg. *Scientific Aspects of the Race Problem.* Washington, D.C.: The Catholic University of America Press, 1941.

Jennings, Herbert Spencer. "Diverse Ideals and Divergent Conclusions in the Study of Behavior of the Lower Organisms." *American Journal of Psychology* 21 (1910): 349–370.

Jennings, Herbert Spencer. "Observed Changes in Hereditary Characters in Relation to Evolution." *Lectures on Heredity.* Washington, D. C.: Washington Academy of Sciences, 1917, 281–301.

Jennings, Herbert Spencer. *Behavior of the Lower Organisms.* New York: Columbia University Press, 1906.

Jennings, Herbert Spencer. *Life and Death, Heredity and Evolution in Unicellular Organisms.* Boston: The Gorham Press, 1920.

Jennings, Herbert Spencer. *Some Implications of Emergent Evolution.* Hanover, New Hampshire: The Sociological Press, 1927.

Jennings, Herbert Spencer. *The Biological Basis of Human Nature.* New York: Norton, 1930.

Jennings, Herbert Spencer. *The Universe and Life*. New Haven: Yale University Press, 1933.

Johnston, John Black. *The Liberal College in a Changing Society*. New York: Century, 1930.

Johnston, John Black. *The Nervous System of Vertebrates*. Philadelphia: P. Blakiston's Son, 1906.

Kappers, C. U. Ariëns, G. Carl Huber, and Elizabeth Caroline Crosby. *The Comparative Anatomy of the Nervous System of Vertebrates, Including Man*. New York: MacMillan, 1936.

Kappers, C. U. Ariëns. "An Appreciation of the Journal of Comparative Neurology at Its Golden Jubilee." *Journal of Comparative Neurology* 74 (1941): 261–298.

Kappers, C. U. Ariëns. "Further Contributions on Neurobiotaxis." *Journal of Comparative Neurology* 27.3 (1917): 261–298.

Kappers, C. U. Ariëns. "The Brain of a Lebanese." *Journal of Comparative Neurology* 56 (1932): 15–26.

Kawin, Ethel. *Children of Preschool Age: Studies in Socio-Economic Status, Social Adjustment and Mental Ability, With Illustrative Cases*. Chicago: University of Chicago Press, 1934.

Kay, Lily. *The Molecular Vision of Life: Caltech, the Rockefeller Foundation and the Rise of the New Biology*. New York: Oxford University Press, 1993.

Keller, Evelyn Fox. *Reflections on Gender and Science*. New Haven: Yale University Press, 1985.

Kennedy, Foster. "Frederick Tilney, M.D.: An Appreciation." *Journal of Nervous and Mental Diseases* 89 (1939): 265–268.

Kevles, Daniel J. "Testing the Army's Intelligence: Psychologists and the Military in World War I." *Journal of American History* 55 (1968): 565–581.

Kevles, Daniel J. *In the Name of Eugenics: Genetics and the Uses of Human Heredity*. New York: Knopf, 1985.

Kingsland, Sharon E. "A Humanistic Science: Charles Judson Herrick and the Struggle for Psychobiology at the University of Chicago." *Perspectives on Science* 1 (1993): 445–477.

Kingsland, Sharon E. "A Man Out of Place: Herbert Spencer Jennings at Johns Hopkins, 1906–1938." *American Zoologist* 27 (1987): 807–817.

Klüver, Heinrich. "Psychology at the Beginning of World War Two: Meditations on the Impending Dismemberment of Psychology Written in 1942." *Journal of Psychology* 28 (1949): 393–410.

Klüver, Heinrich. *Behavior Mechanisms in Monkeys*. Chicago: University of Chicago Press, 1933.

Koffka, Kurt. "Perception: An Introduction to the *Gestalt-Theorie*." *Psychological Bulletin* 19 (1922): 531–585.

Kohler, Robert E. "Science, Foundations and American Universities in the 1920s." *Osiris*, second series 3 (1987): 135–64.

Kohler, Robert E. "The Management of Science: The Experience of Warren Weaver and the Rockefeller Foundation Program in Molecular Biology." *Minerva* 14 (1976): 279–306.

Kohler, Robert E. *Partners in Science: Foundations and Natural Scientists, 1900–45*. Chicago: University of Chicago Press, 1991.

Köhler, Wolfgang. *Gestalt Psychology*. New York: Horace Liveright, 1929.

Köhler, Wolfgang. *The Place of Value in a World of Facts*. New York: Meridian Books, 1938.

Krueger, Robert G. and Clark L. Hull. "An Electro-Chemical Parallel to the Conditioned Reflex." *Journal of General Psychology* 5 (1931): 262–269.

Kubie, Lawrence S. "The Fallacious Use of Quantitative Concepts in Dynamic Psychology." *Psychoanalytic Quarterly* 16 (1947): 507–518.

Lagemann, Ellen Condliffe. *The Politics of Knowledge: The Carnegie Corporation, Philanthropy and Public Policy.* Middletown: Wesleyan University Press, 1989.

Landis, Carney and M. Marjorie Bolles. *Personality and Sexuality of the Physically Handicapped Woman.* New York: P.B. Hoeber, 1942.

Landis, Carney. "Studies of Emotional Reactions I: A Preliminary Study of Facial Expression." *Journal of Experimental Psychology* 7 (October 1924): 325–341.

Landis, Carney. "Studies of Emotional Reactions II: General Behavior and Facial Expression." *Journal of Comparative Psychology* 4 (October-December 1924): 447–509.

Lapiere, Richard T. and Paul R. Farnsworth. *Social Psychology.* New York: McGraw Hill, 1936.

Levy, Steven. "Dr. Edelman's Brain." *The New Yorker,* May 2, 1994: 62–73.

Leys, Ruth and Rand B. Evans. *Defining American Psychology: Correspondence Between Adolf Meyer and Edward Bradford Titchener.* Baltimore: Johns Hopkins University Press, 1990.

Leys, Ruth. "Meyer, Watson and the Dangers of Behaviorism." *Journal of the History of the Behavioral Sciences* 20 (1984): 128–151.

Lipsitt, Lewis P. "Myrtle B. McGraw (1899–1988)." *American Psychologist* 45 (1990): 977.

Ludmerer, Kenneth M. *Genetics and American Society: A Historical Appraisal.* Baltimore: Johns Hopkins University Press, 1972.

Marchand, C. Roland. *The American Peace Movement and Social Reform, 1891–1918.* Princeton: Princeton University Press, 1972.

Marshall, W. H., Clinton N. Woolsey, and Philip Bard. "Observations on Cortical Somatic Sensory Mechanisms of Cat and Monkey." *Journal of Neurophysiology* 4 (1941): 1–24.

May, Henry. *The End of American Innocence: A Study of the First Years of Our Own Time, 1912–1917.* New York: Columbia, 1993.

May, Mark A. "A Retrospective View of the Institute of Human Relations at Yale." *Behavior Science Notes* 6 (1971): 141–172.

May, Mark. *Toward a Science of Human Behavior: A Survey of the Work of the Institute of Human Relations Through Two Decades, 1929–1949.* New Haven: Yale University Press, 1950.

Mayrhauser, Richard von. "Making Intelligence Functional: Walter Dill Scott and Applied Psychological Testing in World War I." *Journal of the History of the Behavioral Sciences* 25 (1989): 60–72.

McGraw, Myrtle B. "A Comparative Study of a Group of Southern White and Negro Infants." *Genetic Psychology Monographs* 10 (1931): 1–104.

McGraw, Myrtle B. "Memories, Deliberate Recall and Speculations." *American Psychologist* 45 (1990): 934–7.

McMahon, A. Michal. "An American Courtship: Psychologists and Advertising Theory in the Progressive Era." *American Studies* 13 (1972): 5–18.

Mettler, Fred A. "Cerebral Function and Cortical Localization." *Journal of General Psychology* 13 (1935): 367–396.

Meyer, Adolf. *Psychobiology: A Science of Man.* (The Thomas W. Salmon Lectures at the New York Academy of Medicine, 1931.) Springfield: Charles C. Thomas, 1957.

Miller, Neal E. and John Dollard. *Social Learning and Imitation.* New Haven: Yale University Press, 1941.

Mitman, Gregg. *The State of Nature: Ecology, Community and American Social Thought, 1900–1950.* Chicago: University of Chicago Press, 1992.

Mitman, Gregg and Anne Fausto-Sterling. "Whatever Happened to Planaria? C. M. Child and the Physiology of Inheritance." *The Right Tools for the Job: At Work in Twentieth Century Life Sciences,* eds. Adele E. Clarke and Joan Fujimura. Princeton: Princeton University Press, 1992, 172–197.

Morawski, J. G. "Organizing Knowledge and Behavior at Yale's Institute of Human Relations." *Isis* 77 (1986): 219–242.

Morawski, J. G., ed. *The Rise of Experimentation in American Psychology.* New Haven: Yale University Press, 1988.

Morgan, Clifford T. "The Hoarding Instinct." *Psychological Review* 54 (1947): 335–341.

Mowry, George. *The Era of Theodore Roosevelt.* New York: Harper, 1958.

Mowry, George. *The Progressive Era, 1900–1918: Recent Literature and New Ideas.* Washington, D. C.: Service Center for Teachers of History, 1964.

Murchison, Carl. *Social Psychology: The Psychology of Political Domination.* Worcester: Clark University Press, 1929.

Murdock, George Peter. *Social Structure.* New York: MacMillan, 1949.

Murphy, Gardner and Lois Barclay Murphy. *Experimental Social Psychology.* New York: Harper Brothers, 1931.

Nelson, Daniel W. *Frederick W. Taylor and the Rise of Scientific Management.* Madison: University of Wisconsin Press, 1980.

O'Donnell, John M. *The Origins of Behaviorism: American Psychology, 1870–1920.* New York: New York University Press, 1985.

O'Leary, J. L. and G. H. Bishop. "Charles Judson Herrick and the Founding of Comparative Neurology." *Archives of Neurology* 3 (1960): 725–731.

Parascandola, John. "Organismic and Holistic Concepts in the Thought of L. J. Henderson." *Journal of the History of Biology* 4 (1971): 63–113.

Pauly, Philip J. "The Loeb-Jennings Debate and the Science of Animal Behavior." *Journal of the History of the Behavioral Sciences* 17 (1981): 504–515.

Pauly, Philip J. "G. Stanley Hall and His Successors: A History of the First Half-Century of Psychology at Johns Hopkins." *One Hundred Years of Psychological Research in America: G. Stanley Hall and the Johns Hopkins Tradition,* eds. Stewart Hulse and Bert F. Green. Baltimore: The Johns Hopkins University Press, 1986.

Pauly, Philip J. *Controlling Life: Jacques Loeb and the Engineering Ideal in Biology.* New York: Oxford University Press, 1987.

Pavlov, Ivan Petrovich. "The Reply of a Physiologist to Psychologists." *Psychological Review* 39 (1932): 91–127.

Pickens, Donald. *Eugenics and the Progressives.* Nashville: Vanderbilt University Press, 1968.

Provine, William B. "Geneticists and Race." *American Zoologist* 26 (1986): 857–887.

Rainger, Ronald, Keith Benson, and Jane Maienschein. *The American Development of Biology.* Philadelphia: University of Pennsylvania Press, 1988.

Robinson, Forrest G. *Love's Story Told: A Life of Henry A. Murray.* Cambridge: Harvard University Press, 1992.

Rockefeller Foundation Annual Report. New York: The Rockefeller Foundation, 1931.

Rodgers, Daniel T. "In Search of Progressivism." *Reviews in American History* 10 (1982): 113–132.

Roofe, Paul G. "Charles Judson Herrick." *Anatomical Record* 137 (1960): 162–4.

Roofe, Paul G. "Neurology Comes of Age: The Herrick-Meyer Scientific Papers Initiating a More Rational Approach to Neurology and Psychobiology." *Journal of the Kansas Medical Society* 64 (1963): 124–9.

Roofe, Paul G. "Some Letters from the Herrick-Lashley Correspondence." *Neuropsychologia* 8 (1970): 3–12.

Rose, Jerzy E. and Clinton Woolsey. "Structure and Relations of Limbic Cortex and Anterior Thalamic Nuclei in Rabbit and Cat." *Journal of Comparative Neurology* 89 (1948): 278–347.

Ross, Dorothy. *The Origins of American Social Science.* Cambridge: Cambridge University Press, 1991.

Rossiter, Margaret W. *Women Scientists in America: Struggles and Strategies to 1940.* Baltimore: Johns Hopkins University Press, 1982.

Sacks, Oliver. "Neurology and the Soul." *The New York Review of Books,* November 22, 1990: 44–50.

Samelson, Franz. "From 'Race Psychology' to 'Studies in Prejudice': Some Observations on the Thematic Reversal in Social Psychology." *Journal of the History of the Behavioral Sciences* 14 (1978): 265–278.

Samelson, Franz. "Organizing for the Kingdom of Behavior: Academic Battles and Organizational Policies in the 20s." *Journal of the History of the Behavioral Sciences* 21 (1985): 33–47.

Samelson, Franz. "The APA Between the World Wars: 1918–1941." *The American Psychological Association: A Historical Perspective,* eds. Rand B. Evans, Virginia Staudt Sexton, and Thomas C. Cadwallader. Washington, D.C.: The American Psychological Association, 1992.

Samelson, Franz. "The Struggle for Scientific Authority: The Reception of Watson's Behaviorism, 1913–1920." *Journal of the History of the Behavioral Sciences* 17 (1981): 399–425.

Samelson, Franz. "World War I Intelligence Testing and the Development of Psychology." *Journal of the History of the Behavioral Sciences* 13 (1977): 274–282.

Schiller, Claire H., trans. and ed. *Instinctive Behavior: The Development of a Modern Concept.* New York: International Universities Press, 1957.

Science Service. "Science at the Century of Progress Exposition." *Science* 77 (1933): 8–10.

Scott, John Paul, and John L. Fuller. "Research on Genetics and Social Behavior at the Roscoe B. Jackson Memorial Laboratory, 1946–1951: A Progress Report." *Journal of Heredity* 42 (1951): 191–196.

Searle, John. *The Rediscovery of the Mind.* Cambridge: MIT Press, 1991.

Shapin, Steven and Simon Schaffer. *Leviathan and the Air-Pump: Hobbes, Boyle and the Experimental Life.* Princeton: Princeton University Press, 1985.

Shapin, Steven. "The Politics of Observation: Cerebral Anatomy and Social Interests in the Edinburgh Phrenology Disputes." *On the Margins of Science: The Social Construction of Rejected Knowledge,* ed. Roy Wallis. Keele: University of Keele, 1979, 139–178.

Shaw, Clifford Robe and Henry D. McKay. *Juvenile Delinquency and Urban Areas: A Study of Rates of Delinquency in Relation to Differential Characteristics of Local Communities in American Cities.* Chicago: University of Chicago Press, 1942.

Shaw, Clifford Robe. *Brothers in Crime.* Chicago: University of Chicago Press, 1938.

Shaw, Clifford Robe. *Delinquency Areas: A Study of the Geographic Distribution of School Truants, Juvenile Delinquents and Adult Offenders in Chicago.* Chicago: University of Chicago Press, 1929.

Shaw, Clifford Robe. *The Jack Roller: A Delinquent Boy's Own Story.* Chicago: University of Chicago Press, 1930.

Shuey, Audrey M. *The Testing of Negro Intelligence.* Lynchburg: J. P. Bell, 1958.

Simmel, Marianne, ed. *The Reach of Mind: Essays in Memory of Kurt Goldstein.* New York: Springer Publishing Co., 1968.

Skinner, Quentin. "Meaning and Understanding in the History of Ideas." *Meaning and Context: Quentin Skinner and His Critics,* ed. James Tully. Princeton: Princeton University Press, 1988, 29–67.

Smith, Laurence D. *Behaviorism and Logical Positivism: A Reassessment of the Alliance.* Stanford: Stanford University Press, 1986.

Smith, Roger. *Inhibition: History and Meaning in the Sciences of Mind and Brain.* Berkeley: University of California Press, 1992.

Sokal, Michael M. "The Gestalt Psychologists in Behaviorist America." *American Historical Review* 89 (1984): 1240–1263.

Sokal, Michael M. "The Origins of the Psychological Corporation." *Journal of the History of the Behavioral Sciences* 17 (1981): 54–67.

Sokal, Michael M. and Patrice Rafail, compilers. *A Guide to Manuscript Collections in the History of Psychology and Related Areas.* Millwood: Kraus International, 1982.

Sokal, Michael M., ed. *An Education in Psychology: James McKeen Cattell's Journal and Letters from Germany and England, 1880–1888.* Cambridge: MIT Press, 1981.

Sokal, Michael M., ed. *Psychological Testing and American Society, 1890–1930.* New Brunswick: Rutgers University Press, 1987.

Sonneborn, Tracy Morton, ed. *The Control of Human Heredity and Evolution.* New York: MacMillan, 1965.

Spearman, Charles Edward. "C. Spearman." *A History of Psychology in Autobiography,* Vol. 1. Worcester: Clark University Press, 1930.

Spearman, Charles Edward. *Creative Mind.* New York: D. Appleton, 1931.

Spearman, Charles Edward. *The Abilities of Man: Their Nature and Measurement.* New York: MacMillan, 1927.

Spearman, Charles Edward. *The Nature of "Intelligence" and the Principles of Cognition.* London: MacMillan, 1923.

Stockard, Charles Rupert. *The Genetic and Endocrinic Basis for Differences in Form and Behavior, As Elucidated by Studies of Contrasted Pure-Line Breeds and Their Hybrids.* American Anatomical Memoirs 19. Philadelphia: Wistar Institute, October 1941.

Stockard, Charles Rupert. *The Physical Basis of Personality.* New York: Norton, 1931.

Stocking, George W., Jr. *Race, Culture and Evolution: Essays in the History of Anthropology.* New York: Free Press, 1986.

Stone, Calvin P. "Methodological Resources for the Experimental Study of Innate Behavior as Related to Environmental Factors." *Psychological Review* 54 (1947): 342–347.

Sturdy, Steve. "Biology as Social Theory: John Scott Haldane and Physiological Regulation." *British Journal for the History of Science* 21 (1988): 315–340.

Swazey, Judith P. "Action Propre et Action Commune." *Journal of the History of Biology* 3 (1970): 213–234.

Taylor, Frederick Winslow. *The Principles of Scientific Management.* New York: Harper and Brothers, 1911.

Thorndike, Edward L. *Human Learning.* New York, Century, 1931.

Thorndike, Edward L., E. O. Bregman, M. N. Cobb, and E. Woodyard. *The Measurement of Intelligence.* New York: Columbia, 1927.

Thurstone, Louis Leon. *Primary Mental Abilities.* Chicago: University of Chicago Press, 1938.

Thurstone, Louis Leon. *The Nature of Intelligence*. New York: Harcourt, Brace and Co., 1924.

Tilney, Frederick and Joshua Rosett. "Brain Lipoids as an Index of Brain Development." *Bulletin of the Neurological Institute of New York* 1 (1931): 28–71.

Tilney, Frederick and Lawrence S. Kubie. "Behavior in Its Relation to the Development of the Brain." Part I, *Bulletin of the Neurological Institute of New York* 1 (1931): 229–313; Part II, *Bulletin of the Neurological Institute of New York* 3 (1933): 252–358.

Tilney, Frederick. *The Master of Destiny: A Biography of the Brain*. Garden City: Doubleday, Doran and Co., 1930.

Tizard, Barbara. "Theories of Brain Localization from Flourens to Lashley." *Medical History* 3 (1959): 132–45.

Tolman, Edward C. "A New Formula for Behaviorism." *Psychological Review* 29 (1922): 44–53.

Triplet, Rodney G. "Harvard Psychology, the Psychological Clinic and Henry A. Murray: A Case Study in the Establishment of Disciplinary Boundaries." *Science at Harvard University: Historical Perspectives,* eds. Clark A. Elliott and Margaret W. Rossiter. Bethlehem, Pa.: Lehigh University Press, 1992, 223–250.

UNESCO. "UNESCO Statement on the Nature of Race and Race Differences." Paris, June 1951. Reprinted in Provine, "Geneticists and Race," *American Zoologist* 26 (1986): 883–886.

Watson, John Broadus. "Is Thinking Merely the Action of the Language Mechanisms?" *British Journal of Psychology* 11 (1921): 87–104.

Watson, John Broadus. "John B. Watson." *A History of Psychology in Autobiography,* Vol. 3, ed. Carl Murchison. Worcester: Clark University Press, 1936.

Watson, John Broadus. "Psychology as the Behaviorist Views It." *Psychological Review* 20 (1913): 158–178.

Watson, John Broadus. *Behavior: An Introduction to Comparative Psychology*. New York: Henry Holt, 1914.

Watson, John Broadus. *Behaviorism*. New York: Norton, 1924.

Watson, John Broadus. *Psychology from the Standpoint of a Behaviorist*. Philadelphia: J. B. Lippincott, 1919.

Wiebe, Robert H. *The Search for Order, 1870–1920*. New York: Hill and Wang, 1977.

Worder, Frederic G., Judith P. Swazey, and George Adelman. *The Neurosciences: Paths of Discovery*. Cambridge: MIT Press, 1975.

Yerkes, Robert Mearns. "Man-Power and Military Effectiveness: The Case for Human Engineering." *Journal of Consulting Psychology* 5 (1941): 205–209.

Yerkes, Robert Mearns. "Psychology and Defense." *Proceedings of the American Philosophical Society* 84 (1941): 527–542.

Yerkes, Robert Mearns. "Robert Mearns Yerkes, Psychobiologist." *A History of Psychology in Autobiography,* Vol. 2, ed. Carl Murchison. Worcester: Clark University Press, 1932.

Yerkes, Robert Mearns. "Yale Laboratories of Comparative Psychobiology." *Comparative Psychology Monographs* 8 (1932): 1–33.

Yerkes, Robert Mearns. *Chimpanzees: A Laboratory Colony*. New Haven: Yale University Press, 1943.

Young, Robert M. *Mind, Brain and Adaptation in the Nineteenth Century*. New York: Oxford University Press, 1970.

Zinn, Earl. "History, Purpose and Policy of the National Research Council's Committee for Research on Problems of Sex." *Mental Hygiene* 8 (1924): 94–105.

Index

Aberle, Sophie D., 151–2
ability, determination of *(see also* intelligence), 73–5, 77–8, 82, 83, 129
ablation, of areas in cortex, 10, 51, 55–60, 61f, 67–8, 112, 189
Addams, Jane, 89, 92
Adler, Herman, 116–17
Adrian, E. D., 84
Allport, Gordon, 144
American Neurological Association, 183
American Psychological Association, 21, 68–70, 75, 133, 171
Angell, James Rowland, 32, 125–26
anthropoid apes, scientific study of, 154–59
anthropology, 125, 126, 131, 176, 179, 181n23, 182
aphasia, 6–7, 51
architectonics (brain mapping), 80–85
attention, 138, 189

bacteriology, 18, 19, 20
Baernstein, H. D., 130
Bain, Alexander, 6
Baker, Edith Ann, *see* Lashley, Edith Ann Baker
Baldwin, James Mark, 32
Beach, Frank: on Clark Hull, 122; on role of heredity in behavior, 112, 119; on sexual behavior, 112, 152, 190; as student of Lashley, 18, 19, 112, 119, 152, 186, 190
behavior *(see also* behaviorism; intelligence; learning; reflex): definitions of, 25, 39, 124; in determining organismic structure, 90–92, 94–96; of protozoa, 26–27
behavior genetics, 16, 176, 182–86, 188
Behavior Mechanisms in Monkeys (Klüver), 118
Behavior of the Lower Organisms (Jennings), 25, 27
Behavior Research Fund, 49, 109, 116–18, 186
behaviorism, 12–13, 15, 23, 88, 101, 105, 113, 122, 141, 143, 145, 164, 188, 189; and biology, 36, 39, 41–42, 124, 130–31; Hull's

(neo-behaviorism), 120–24, 126, 128–32, 134–38; Lashley's, 34–35, 37–38, 40–47, 48, 50–56, 61, 64–68, 71; and neurology, 13, 35, 41–42, 51–55, 56, 63–64, 124, 128, 129, 132, 135; practical applications of, 42; Watson's, 32–40, 63–64
Bell, Charles, 6
Bell Curve, The (Herrnstein and Murray), 190–91
Bentley, Madison, 24n18
biology, *see* psychology, relation to biology
Bird Key, *see* Dry Tortugas
Bishop, George H., 147–49, 163–64, 170
Boring, Edwin G., 11, 14, 69–70, 143–44
brain anatomy, *see* architectonics; cortex; localization; neurology
brain function, *see* learning; localization; reflex
Broca, Paul, 5–6, 7, 8, 9
Brodmann, K., 80
Bronner, Augusta Fox, 93
Brown University, 94
Bruce, Darryl, 18–19, 23, 28, 30, 47
Bureau of Social Hygiene, 149–50
Burnham, John, 89

Cannon, Walter B., 149
Carmichael, Leonard, 119
Carnegie Corporation, 155
Carr, Harvey, 24n18
Chicago, University of: Lashley at, 15, 105, 139, 143–44; psychology at, 32, 49, 85, 125, 139
Child, Charles Manning: collaboration with C. J. Herrick, 103; gradient theory of, 90–93, 105–8; on individuality, 97, 106; influence on Lashley, 105–8; as member of American school of psychobiology, 15, 89; organism-society analogy of, 49, 100; and Progressive reform, 92–93
Clark, George, 80–82, 85
Clinic of Child Development, 97, 125
Coghill, George Ellett, 89, 93–97, 99, 103
cognitive science, 3–4, 14, 15, 189

213